Chapter 1: Introduction to Vasopressin and Its Role in Physiology

Vasopressin, also known as antidiuretic hormone (ADH), is a peptide hormone that plays a crucial role in maintaining the body's internal balance of water, electrolytes, and blood pressure. Secreted primarily by the hypothalamus and stored in the posterior pituitary gland, vasopressin is involved in regulating vital processes such as fluid homeostasis, vascular tone, and central nervous system function. Despite its relatively small size and straightforward structure, vasopressin has a profound impact on human physiology, affecting multiple organ systems and influencing both short- and long-term bodily functions.

The hormone's primary action is to conserve water within the kidneys, concentrating urine while preventing excessive water loss. However, vasopressin's effects extend beyond the kidneys, as it also plays a significant role in regulating blood pressure and responding to various stressors. The hormone achieves these effects by binding to specific receptors, known as vasopressin receptors, located in target tissues such as the kidneys, blood vessels, and brain. Vasopressin agonists—synthetic compounds that mimic the action of vasopressin—have been developed to harness these mechanisms for therapeutic purposes, treating a variety of conditions related to cardiovascular health, diabetes insipidus, shock, and more.

In this chapter, we will explore the physiological basis of vasopressin's actions, the different conditions that may lead to vasopressin dysregulation, and how these insights paved the way for the development of vasopressin agonists. We will cover the following key concepts:

1.1. The Physiological Role of Vasopressin

Vasopressin's role in the human body is multifaceted. The hormone's primary function is to help the body retain water and maintain osmotic balance. It achieves this by acting on the kidneys, increasing water reabsorption from the urine back into the bloodstream, thereby concentrating urine and reducing urine output. However, its effects are not limited to the kidneys. Vasopressin also influences vascular smooth muscle contraction, helping to maintain blood pressure and perfusion of critical organs, particularly during times of stress or blood loss.

Water Homeostasis

The most widely known action of vasopressin is its role in regulating water balance. The hormone responds to signals that indicate dehydration or increased plasma osmolality, typically in the form of low blood volume or high sodium concentration. When vasopressin is released from the posterior pituitary, it binds to V2 receptors in the kidneys, promoting the insertion of water channels (aquaporins) into the walls of the collecting ducts. This process facilitates the reabsorption of water, reducing urine output and helping the body conserve fluid. This regulation is critical for maintaining proper hydration and electrolyte balance.

Vasoconstriction and Blood Pressure Regulation

In addition to its renal effects, vasopressin has a vasoconstrictor action through the V1a receptors located in vascular smooth muscle. When vasopressin binds to these receptors, it triggers intracellular signaling pathways that result in the contraction of blood vessels, which increases systemic vascular resistance and elevates blood pressure. This vasoconstrictor effect is particularly important during conditions like hemorrhagic shock, where maintaining blood pressure is crucial for organ perfusion.

Neuroendocrine and Behavioral Effects

Vasopressin also plays a role in the central nervous system, influencing behavior and stress responses. In the brain, vasopressin acts on various areas such as the hypothalamus, amygdala, and pituitary gland, where it modulates processes like social behavior, aggression, and stress regulation. Vasopressin has been implicated in the regulation of the hypothalamic-pituitary-adrenal (HPA) axis, a key stress response system. Research has suggested that vasopressin may contribute to feelings of anxiety and bonding, particularly in the context of social and reproductive behavior.

1.2. The Structure and Secretion of Vasopressin

Vasopressin is a peptide hormone composed of nine amino acids, and it is synthesized in the hypothalamus as a larger precursor molecule known as preprovasopressin. This precursor is cleaved to form the mature vasopressin molecule, which is then transported to the posterior pituitary gland for storage and eventual release into the bloodstream. The secretion of vasopressin is tightly regulated by osmotic and volume sensors located in the hypothalamus, which respond to changes in blood osmolality (the concentration of solutes such as sodium in the blood) and blood volume.

The primary triggers for vasopressin release include:

- **Increased plasma osmolality** (detected by osmoreceptors in the hypothalamus) signaling the need to conserve water.

- **Decreased blood volume** (detected by baroreceptors in the heart and blood vessels) signaling the need to increase blood pressure and volume.

- **Stress**: Both physical and psychological stressors can stimulate the release of vasopressin, potentially increasing blood pressure and aiding the body's response to a perceived threat.

1.3. Vasopressin Receptors

Vasopressin exerts its effects through specific receptor subtypes, each associated with different tissues and physiological outcomes. The three main vasopressin receptors are:

- **V1a Receptors**: Found in vascular smooth muscle, the V1a receptor mediates vasoconstriction and plays a key role in regulating blood pressure.

- **V1b (V3) Receptors**: Located primarily in the anterior pituitary gland, the V1b receptors help regulate the release of ACTH (adrenocorticotropic hormone), which influences cortisol production and the body's stress response.

- **V2 Receptors**: Predominantly expressed in the kidneys, particularly in the collecting ducts, V2 receptors facilitate water reabsorption through the action of aquaporins, contributing to water homeostasis.

These receptors have distinct roles, and vasopressin agonists can be designed to selectively target one receptor subtype over others, providing therapeutic benefits in specific clinical scenarios.

1.4. Pathophysiological Conditions Involving Vasopressin

Dysregulation of vasopressin secretion or receptor activity can lead to a variety of pathophysiological conditions:

- **Diabetes Insipidus (DI)**: A condition characterized by excessive urination and thirst, either due to a deficiency of vasopressin (central DI) or resistance to its effects in the kidneys (nephrogenic DI).
- **Hyponatremia**: Low sodium levels in the blood can occur when vasopressin secretion is inappropriate, leading to excessive water retention and dilution of sodium.
- **Hypotension and Shock**: In conditions such as septic shock, where vasodilation occurs, vasopressin may be used therapeutically to restore vascular tone and increase blood pressure.

Vasopressin agonists are especially useful in these and other conditions, allowing clinicians to correct or modify the body's natural vasopressin activity for therapeutic benefit.

1.5. Vasopressin Agonists: An Overview

Vasopressin agonists are synthetic compounds that mimic the action of endogenous vasopressin, typically by binding to vasopressin receptors and activating the same biological pathways. These agents have proven to be invaluable in clinical practice, particularly in the treatment of shock, vasodilatory hypotension, diabetes insipidus, and certain forms of bleeding. Synthetic vasopressin agonists, such as desmopressin and terlipressin, have been developed to target specific vasopressin receptor subtypes and provide more targeted and efficient therapeutic effects.

In subsequent chapters, we will delve deeper into the mechanisms of action, clinical applications, and emerging research on vasopressin agonists. However, it is essential to first understand the physiological role of vasopressin and its receptor interactions, as these form the foundation for all clinical applications of vasopressin agonists.

Conclusion

Vasopressin is a key regulator of fluid balance, blood pressure, and stress responses in the human body. Its ability to act on multiple organ systems, including the kidneys, blood vessels, and brain, makes it an essential component of homeostasis. Vasopressin agonists—mimicking the natural hormone—have become an indispensable tool in modern medicine, particularly in the management of critical conditions such as shock, diabetes insipidus, and vasodilatory hypotension. Understanding the physiological and biochemical principles that govern vasopressin's actions is the first step in mastering the clinical use of vasopressin agonists. As we explore this fascinating area of pharmacology, we will uncover how these agents are used to treat a variety of conditions, with a focus on optimizing their application to enhance patient outcomes.

Chapter 2: The Science Behind Vasopressin Agonists: A Biochemical Overview

Vasopressin agonists are synthetic compounds designed to replicate or enhance the actions of the natural vasopressin hormone, targeting the body's vasopressin receptors to promote specific physiological effects. Understanding the biochemical principles behind vasopressin agonists is essential for mastering their clinical applications. This chapter explores the chemical structure of vasopressin agonists, how they interact with the vasopressin receptors, their development, and the specific molecular mechanisms by which they exert their effects.

2.1. The Chemical Structure of Vasopressin Agonists

Vasopressin is a nonapeptide—a molecule composed of nine amino acids—and its molecular structure is a critical factor in its ability to bind to vasopressin receptors. The structure of synthetic vasopressin agonists is often modeled after the natural hormone, but with modifications that enhance stability, receptor selectivity, and duration of action.

Peptide Backbone and Modifications

Like endogenous vasopressin, most vasopressin agonists are peptide-based. These molecules typically have a similar backbone structure, with a series of amino acids linked by peptide bonds. However, synthetic agonists often feature modifications that increase their resistance to enzymatic degradation, extend their half-life, or allow them to selectively target one of the three vasopressin receptor subtypes (V1, V2, or V3).

For example, **desmopressin** (DDAVP), a widely used synthetic vasopressin agonist, is a modified form of vasopressin where the arginine residue at position 8 is replaced with a desmethyl group. This alteration improves its selectivity for the V2 receptor in the kidneys, enhancing its antidiuretic effect while minimizing vasoconstrictor activity.

Analogs and Derivatives

Beyond desmopressin, other vasopressin agonists have been developed with specific structural modifications aimed at enhancing receptor selectivity or targeting specific therapeutic needs. These include:

- **Terlipressin**, a prodrug of lysine-vasopressin, which has a longer duration of action and greater affinity for the V1 receptors, making it effective in managing septic shock and variceal bleeding.
- **Felypressin**, another synthetic analog, is primarily used in dentistry and other minor surgical procedures due to its potent vasoconstrictor effects.

These synthetic modifications allow for greater precision in targeting the therapeutic actions of vasopressin, making vasopressin agonists useful for a broad range of clinical indications.

2.2. Binding to Vasopressin Receptors

The effectiveness of vasopressin agonists hinges on their ability to bind to and activate the appropriate vasopressin receptor. As discussed in Chapter 1, there are three primary types of vasopressin receptors—V1a, V1b (V3), and V2—each of which is located in different tissues and mediates distinct physiological responses.

V1a Receptors: These are predominantly found in the vascular smooth muscle and are responsible for mediating vasoconstriction. Vasopressin agonists that target the V1a receptor are often used in shock management and blood pressure regulation. The binding of vasopressin to V1a receptors activates phospholipase C (PLC), leading to the formation of inositol trisphosphate (IP3) and diacylglycerol (DAG), which trigger intracellular calcium release and cause smooth muscle contraction. This vasoconstriction increases systemic vascular resistance and blood pressure.

V2 Receptors: Located in the kidneys, V2 receptors are responsible for promoting water reabsorption from the urine back into the bloodstream. Vasopressin agonists that primarily target V2 receptors, like desmopressin, increase the insertion of aquaporin-2 channels in the collecting ducts of the kidneys, thereby enhancing water retention. This effect is crucial in the treatment of conditions such as diabetes insipidus, where the kidneys fail to retain water properly.

V3 Receptors: Found mainly in the anterior pituitary gland, V3 receptors are involved in the regulation of the release of adrenocorticotropic hormone (ACTH), which stimulates cortisol production from the adrenal glands. While less commonly targeted by vasopressin agonists, the V3 receptor still plays an important role in the body's stress response.

The selective binding of vasopressin agonists to these receptors enables precise modulation of various physiological processes. By designing agonists with affinity for specific receptor subtypes, researchers and clinicians can optimize the therapeutic effects while minimizing undesirable side effects.

2.3. Mechanisms of Action

Once a vasopressin agonist binds to its target receptor, it initiates a series of intracellular events that culminate in a physiological response. The key mechanisms of action depend on the receptor subtype involved and the tissue in which it is expressed.

V1a Receptor Activation

When a vasopressin agonist binds to the V1a receptor, it activates a G-protein-coupled receptor (GPCR) signaling pathway. This pathway stimulates the activation of phospholipase C (PLC), which in turn increases the levels of intracellular second messengers, inositol trisphosphate (IP3) and diacylglycerol (DAG). These molecules mobilize calcium ions from intracellular stores, resulting in vasoconstriction of smooth muscle cells. This action is critical in maintaining blood pressure, particularly in situations where vasodilation leads to hypotension, such as in septic shock.

V2 Receptor Activation

Vasopressin agonists that target the V2 receptor primarily influence water homeostasis by increasing water reabsorption in the kidneys. When the agonist binds to V2 receptors on the cells of the renal collecting ducts, it activates the adenylate cyclase pathway, increasing cyclic AMP (cAMP) levels. Elevated cAMP activates protein kinase A (PKA), which subsequently promotes the insertion of aquaporin-2 (AQP2) water channels into the apical membrane of collecting duct cells. These channels allow water to be reabsorbed from the urine back into the bloodstream, leading to reduced urine output and increased fluid retention.

V3 Receptor Activation

While the V3 receptor is less commonly targeted, it plays an important role in regulating the body's response to stress. The activation of V3 receptors in the pituitary gland can influence the release of ACTH, which in turn stimulates the adrenal glands to produce cortisol. Cortisol is a key hormone in the body's stress response, helping to increase glucose production and suppress inflammation.

2.4. Pharmacological Classification of Vasopressin Agonists

Vasopressin agonists can be broadly categorized based on their receptor selectivity and clinical use. These agents may act as non-selective vasopressin receptor agonists, which activate all three receptor subtypes, or they may be highly selective for one receptor subtype, depending on the intended therapeutic goal.

Non-selective Agonists

Vasopressin

Selective Agonists

- **Desmopressin** is a synthetic analog of vasopressin with high selectivity for the V2 receptor. It is used primarily in the management of diabetes insipidus and certain bleeding disorders. Its V2 receptor specificity reduces the vasoconstrictor effects seen with natural vasopressin, making it suitable for conditions requiring antidiuretic effects without significant blood pressure increases.

- **Terlipressin** is an example of a V1a receptor-selective vasopressin agonist. Its primary use is in the treatment of variceal bleeding and septic shock, where strong vasoconstriction is needed to counteract hypotension.

- **Felypressin**, which has a preference for the V1a receptor, is commonly used in dental procedures as a vasoconstrictor to control bleeding during local anesthesia. Its short duration of action and vasoconstrictive properties make it effective for these short-term applications.

2.5. Clinical Development and Innovations in Vasopressin Agonists

The development of vasopressin agonists has been driven by the need to refine the therapeutic applications of the natural hormone. Early vasopressin therapies were non-selective and could cause a variety of side effects due to the broad activation of all receptor subtypes. Over time, researchers have developed more selective agonists that target specific receptors, allowing for more tailored treatments with fewer adverse effects.

Innovation continues to play a central role in the field, with new forms of vasopressin agonists being designed to have improved pharmacokinetic profiles, such as longer half-lives, faster onset of action, or more potent receptor selectivity. Recent advances have also focused on delivering vasopressin agonists through novel drug delivery systems, such as sustained-release formulations or nasal sprays, which can improve patient compliance and reduce side effects.

2.6. Conclusion

The science behind vasopressin agonists is deeply rooted in the molecular interactions between synthetic peptides and vasopressin receptors. By understanding the chemical structure, receptor binding mechanisms, and pharmacological actions of these agents, clinicians can optimize their use in treating a variety of conditions. The development of selective vasopressin agonists has opened new avenues for targeted therapies, allowing for better control over fluid balance, blood pressure, and other vital processes. In the next chapter, we will explore in-depth the three primary vasopressin receptor subtypes, providing a deeper understanding of their distinct roles in physiology and how they influence clinical outcomes when targeted by vasopressin agonists.

Chapter 3: Vasopressin Receptors: An In-Depth Analysis

Vasopressin agonists exert their therapeutic effects primarily through interactions with specific vasopressin receptors. These receptors are integral to a variety of physiological processes, including blood pressure regulation, water balance, and stress responses. A deeper understanding of vasopressin receptors—their distribution, molecular mechanisms, and the biological responses they mediate—is essential for effectively using vasopressin agonists in clinical settings.

This chapter provides an in-depth analysis of the three main vasopressin receptor subtypes—V1a, V1b (V3), and V2. We will explore their molecular structure, signaling pathways, tissue distribution, and the physiological processes they influence. We will also discuss how targeting these receptors with vasopressin agonists can provide therapeutic benefits for various medical conditions.

3.1. Overview of Vasopressin Receptors

Vasopressin receptors are a group of G-protein-coupled receptors (GPCRs) that mediate the actions of vasopressin and vasopressin-like agonists. There are three distinct subtypes—V1a, V1b (V3), and V2—each with unique characteristics, tissue localization, and signaling pathways. These receptors play pivotal roles in cardiovascular function, renal water reabsorption, stress response, and fluid homeostasis.

G-Protein-Coupled Receptor (GPCR) Mechanism

Vasopressin receptors belong to the large family of GPCRs, which are involved in transmitting signals from the extracellular environment to the inside of the cell. Upon binding of vasopressin or an agonist, the receptor undergoes a conformational change that activates an intracellular signaling cascade, typically through the activation of G-proteins. These proteins then trigger downstream effects, including the generation of second messengers like inositol trisphosphate (IP3) or cyclic AMP (cAMP), which regulate various cellular processes.

3.2. V1a Receptor: The Vasoconstrictor

The V1a receptor is the primary vasopressin receptor involved in vasoconstriction. It is located predominantly in the vascular smooth muscle, but is also present in the liver, heart, and brain. When activated by vasopressin or a selective agonist, the V1a receptor triggers a series of events that culminate in the contraction of smooth muscle cells, resulting in vasoconstriction and an increase in blood pressure.

Molecular Mechanism of V1a Receptor Activation

Upon binding of vasopressin, the V1a receptor activates a Gq protein, which in turn activates phospholipase C (PLC). PLC catalyzes the breakdown of phosphatidylinositol bisphosphate (PIP2) into two second messengers—inositol trisphosphate (IP3) and diacylglycerol (DAG). These molecules initiate intracellular calcium release and activate protein kinase C (PKC), leading to smooth muscle contraction.

The vasoconstrictive effect of the V1a receptor is essential for the body's ability to regulate blood pressure, particularly in states of hypotension. In critical care settings, vasopressin agonists targeting the V1a receptor can be used to raise blood pressure in patients with shock, including septic shock, hemorrhagic shock, and vasodilatory shock.

Physiological Functions of V1a Activation

- **Vasoconstriction and Blood Pressure Regulation**: The activation of V1a receptors in smooth muscle cells leads to vasoconstriction, which increases systemic vascular resistance and raises blood pressure.
- **Hepatic Glycogenolysis**: In the liver, V1a receptor activation contributes to glycogen breakdown, which increases glucose availability during stress.
- **Central Nervous System**: V1a receptors in the brain are involved in modulating social behavior, emotional responses, and memory. Research has shown that V1a receptor antagonists can influence behaviors related to anxiety and stress.

3.3. V1b Receptor (V3): The Stress Modulator

The V1b receptor, often referred to as V3, is primarily located in the anterior pituitary gland, where it plays a crucial role in regulating the release of adrenocorticotropic hormone (ACTH). ACTH stimulates the adrenal glands to release cortisol, a hormone involved in the body's response to stress. The V1b receptor is also found in the pancreas and the central nervous system, where it can influence various stress-related pathways.

Molecular Mechanism of V1b Receptor Activation

The activation of the V1b receptor involves a Gq-mediated signaling pathway similar to that of the V1a receptor. However, the primary consequence of V1b activation is the release of ACTH from the pituitary gland, rather than vasoconstriction. This release of ACTH initiates the hypothalamic-pituitary-adrenal (HPA) axis, leading to increased cortisol production in response to stress.

Physiological Functions of V1b Activation

- **Stress Response**: V1b receptor activation is key to the HPA axis, which governs the body's physiological response to stress. Increased cortisol levels help regulate metabolism, immune function, and the body's reaction to stress.

- **Regulation of ACTH Secretion**: The release of ACTH stimulates the adrenal cortex to produce cortisol, which in turn helps mobilize energy resources, particularly glucose, during stress.

- **Potential Role in Psychiatric Disorders**: Given its role in stress regulation, the V1b receptor is being investigated as a target for therapeutic interventions in psychiatric conditions such as depression, anxiety, and post-traumatic stress disorder (PTSD).

3.4. V2 Receptor: The Water Retention Regulator

The V2 receptor plays a pivotal role in regulating fluid balance by modulating water reabsorption in the kidneys. Located predominantly in the renal collecting ducts, the V2 receptor is responsible for activating pathways that increase the reabsorption of water into the bloodstream, reducing urine output and conserving body fluids. This function is crucial in maintaining osmotic balance and preventing dehydration.

Molecular Mechanism of V2 Receptor Activation

When vasopressin binds to the V2 receptor, it activates the Gs protein, which stimulates adenylate cyclase. This increases levels of cyclic AMP (cAMP), which activates protein kinase A (PKA). PKA, in turn, promotes the insertion of aquaporin-2 (AQP2) channels into the apical membrane of the renal collecting duct cells. These water channels allow water to pass from the urine back into the bloodstream, thus concentrating the urine and conserving body water.

Physiological Functions of V2 Activation

- **Water Reabsorption in the Kidneys**: The primary function of the V2 receptor is to increase water reabsorption in the kidneys. By promoting the insertion of AQP2 channels, V2 receptor activation ensures that water is retained during periods of dehydration or low fluid intake.

- **Antidiuretic Effect**: In conditions such as diabetes insipidus (DI), where the kidneys fail to respond to vasopressin, synthetic V2-selective agonists like desmopressin are used to mimic the antidiuretic effects of the natural hormone, preventing excessive urination and restoring fluid balance.

3.5. Receptor Crosstalk and Interactions

Although the three vasopressin receptors have distinct functions, there is some level of crosstalk and overlap in their signaling pathways. For example, the V1a and V2 receptors can influence each other's activity, and in some tissues, V1b receptors may modulate the effects of the other two receptor types.

Understanding the complex interactions between vasopressin receptors is essential for developing new pharmacological agents and optimizing existing therapies. Advances in receptor biology and pharmacology will allow for more precise targeting of these receptors, improving the therapeutic outcomes of vasopressin agonists in a variety of clinical contexts.

3.6. Clinical Implications of Vasopressin Receptor Targeting

Vasopressin agonists, which are designed to selectively activate one or more of these receptors, have broad therapeutic potential. By understanding the distribution and molecular mechanisms of vasopressin receptors, clinicians can tailor treatments for specific conditions. For example:

- **V1a-selective agonists** (e.g., terlipressin) are useful in managing shock, particularly in cases of septic or hemorrhagic shock, by increasing vasoconstriction and raising blood pressure.
- **V2-selective agonists** (e.g., desmopressin) are critical in the treatment of diabetes insipidus, where they help concentrate urine and retain water.
- **V1b-targeted therapies** could potentially be used to modulate stress responses and improve outcomes in stress-related psychiatric disorders.

3.7. Conclusion

The vasopressin receptor subtypes—V1a, V1b, and V2—each play distinct and vital roles in regulating blood pressure, fluid balance, and the stress response. A thorough understanding of these receptors and their signaling pathways is crucial for optimizing the use of vasopressin agonists in clinical practice. As research continues to evolve, new receptor-selective therapies will likely emerge, providing more precise and effective treatments for a wide range of medical conditions. In the next chapter, we will delve deeper into the synthesis, regulation, and functions of endogenous vasopressin, exploring how the body naturally produces and controls this essential hormone.

Chapter 4: Endogenous Vasopressin: Synthesis, Regulation, and Function

Endogenous vasopressin, also known as antidiuretic hormone (ADH), is a peptide hormone that plays a critical role in maintaining fluid balance, regulating blood pressure, and supporting the body's response to stress. Produced primarily in the hypothalamus and released by the posterior pituitary gland, vasopressin's effects are mediated through its interactions with specific vasopressin receptors, which are found in various tissues, including the kidneys, blood vessels, and brain.

This chapter will explore the synthesis, regulation, and physiological functions of endogenous vasopressin, providing a foundational understanding of how the body naturally produces and controls this vital hormone. By examining these processes, we can gain insights into how vasopressin agonists can be used to mimic or modulate its effects in clinical settings, enhancing therapeutic outcomes for patients with disorders related to fluid balance, blood pressure regulation, and other related conditions.

4.1. Synthesis and Release of Vasopressin

Vasopressin is synthesized as a precursor protein, preprovasopressin, in the magnocellular neurons of the hypothalamus. These neurons are primarily located in the paraventricular and supraoptic nuclei of the hypothalamus, two key regions involved in the regulation of water balance and blood pressure. The synthesis of vasopressin occurs through a series of steps:

1. **Transcription and Translation**: The gene encoding vasopressin (AVP) is transcribed in the hypothalamic neurons. The resulting mRNA is translated into preprovasopressin, a large precursor protein.

2. **Cleavage of Preprovasopressin**: Preprovasopressin undergoes post-translational modification within the hypothalamus. It is cleaved into several components, including vasopressin, neurophysin II (a carrier protein), and copeptin (a peptide with no known biological function). Neurophysin II binds tightly to vasopressin, preventing its premature release.

3. **Transport and Storage**: The cleaved vasopressin is then transported along axons to the posterior pituitary gland, where it is stored in vesicles until needed.

4. **Release**: When the body detects changes in blood osmolality, blood volume, or blood pressure, signals are sent to the hypothalamus and posterior pituitary gland to release vasopressin into the bloodstream. The primary stimuli for vasopressin release are:

- **Increased Plasma Osmolality**: High levels of sodium or other solutes in the blood stimulate osmoreceptors in the hypothalamus, triggering vasopressin secretion.

- **Decreased Blood Volume or Blood Pressure**: Baroreceptors in the heart and arteries detect a decrease in blood volume or pressure and signal the hypothalamus to release vasopressin.

- **Stress**: Physical or emotional stress can also trigger vasopressin release via the activation of the hypothalamic-pituitary-adrenal (HPA) axis.

4.2. Regulation of Vasopressin Secretion

The release of vasopressin is tightly regulated by both osmotic and non-osmotic signals, ensuring that fluid balance is maintained under a wide variety of physiological conditions. This regulation occurs through several mechanisms:

Osmoregulation

- **Osmoreceptors in the Hypothalamus**: Specialized neurons in the hypothalamus called osmoreceptors detect changes in plasma osmolality. When the osmolality increases (i.e., when the blood becomes more concentrated due to dehydration or high sodium intake), osmoreceptors trigger the release of vasopressin. This action promotes water reabsorption in the kidneys and restores fluid balance by concentrating urine.

- **Negative Feedback Loop**: As vasopressin acts to increase water reabsorption, blood osmolality decreases. When osmolality reaches normal levels, the osmoreceptors sense this change and inhibit further release of vasopressin, creating a feedback loop that maintains homeostasis.

Volume Regulation

- **Baroreceptors**: In addition to osmotic signals, baroreceptors located in the aortic arch and carotid sinus sense changes in blood volume and pressure. A decrease in blood volume or pressure—due to factors like blood loss, dehydration, or hypotension—activates these baroreceptors and triggers vasopressin release. This response serves to constrict blood vessels and increase blood pressure through V1a receptor activation, while simultaneously promoting water retention through V2 receptor activation in the kidneys.

- **Renin-Angiotensin-Aldosterone System (RAAS)**: The RAAS also plays a role in vasopressin regulation. When blood volume or pressure drops, the kidneys release renin, which leads to the production of angiotensin II. Angiotensin II not only stimulates the release of aldosterone (to increase sodium reabsorption) but also potentiates vasopressin secretion. This combined response helps restore both blood volume and osmolality.

4.3. Physiological Functions of Vasopressin

Vasopressin's functions extend beyond just regulating fluid balance and blood pressure. The hormone has a broad range of physiological effects, all of which are crucial for maintaining homeostasis under normal and stress conditions.

Water and Electrolyte Homeostasis

The most well-known function of vasopressin is its ability to regulate water balance by promoting water reabsorption in the kidneys. Through activation of V2 receptors in the renal collecting ducts, vasopressin increases the permeability of the membranes to water, allowing water to be reabsorbed from the urine back into the bloodstream.

Antidiuretic Effect

Blood Pressure Regulation

Vasopressin also plays a critical role in regulating vascular tone and blood pressure. Through activation of V1a receptors in smooth muscle cells, vasopressin causes vasoconstriction, which increases systemic vascular resistance and raises blood pressure. This effect is particularly important in times of stress or blood volume depletion, where vasopressin helps to restore blood pressure and ensure perfusion of vital organs.

Role in Social Behavior and Cognition

In addition to its physiological functions in fluid balance and blood pressure, vasopressin has significant effects on the brain. It has been implicated in modulating behavior, particularly in the context of social bonding, aggression, and stress response. Research has shown that vasopressin levels may influence parental behavior, trust, and social recognition. In fact, it is sometimes referred to as the "social hormone" due to its role in regulating social interactions and emotional responses.

Stress and Adaptation

Vasopressin is released in response to both osmotic and non-osmotic stimuli, including physical and emotional stress. When released as part of the stress response, vasopressin works in concert with other hormones, such as cortisol, to increase alertness, mobilize energy, and regulate blood pressure. This response helps the body adapt to challenging situations by ensuring that sufficient water and blood pressure are maintained during times of physiological strain.

4.4. Disorders Related to Vasopressin Secretion

Disruptions in the synthesis, regulation, or action of vasopressin can lead to a variety of clinical conditions, including:

- **Diabetes Insipidus (DI)**: DI is characterized by excessive urination and thirst due to a deficiency in vasopressin or a resistance to its action. Central DI occurs when there is a lack of vasopressin production in the hypothalamus or pituitary, while nephrogenic DI occurs when the kidneys fail to respond to vasopressin despite normal levels of the hormone.

- **Syndrome of Inappropriate Antidiuretic Hormone (SIADH)**: SIADH is a condition characterized by excessive vasopressin secretion, leading to water retention, hyponatremia (low sodium levels), and dilutional hyposmolality. It is often seen in conditions like lung cancer, head trauma, and certain medications.

- **Hypertension**: Chronic activation of the V1a receptors can contribute to hypertension by promoting vasoconstriction and increasing systemic vascular resistance. Conditions such as essential hypertension or stress-induced hypertension may involve dysregulation of vasopressin release or receptor function.

4.5. Conclusion

Endogenous vasopressin plays a critical role in regulating fluid balance, blood pressure, and stress response. Its synthesis and release are tightly controlled by a combination of osmotic and non-osmotic factors, ensuring that the body can adapt to both hydration status and changes in blood pressure. Disruptions in vasopressin signaling can lead to a range of clinical disorders, underscoring the importance of maintaining balanced vasopressin function. Understanding the physiology of endogenous vasopressin provides essential insights into how vasopressin agonists work and why they are effective in treating conditions such as diabetes insipidus, shock, and other disorders related to fluid balance.

In the next chapter, we will explore the mechanism of action of vasopressin agonists, examining how these agents mimic or enhance the physiological effects of endogenous vasopressin in various clinical contexts.

Chapter 5: Mechanism of Action: How Vasopressin Agonists Work

Vasopressin agonists, a class of drugs designed to mimic or enhance the action of endogenous vasopressin, are widely used in clinical practice to manage a range of conditions, including diabetes insipidus, shock, and cardiovascular disorders. Understanding the mechanism of action of these agents is crucial for their effective use in therapeutic settings. This chapter delves into how vasopressin agonists exert their physiological effects, the role of vasopressin receptors in mediating these actions, and how their pharmacodynamic properties are leveraged in clinical practice.

5.1. Overview of Vasopressin Agonists

Vasopressin agonists are synthetic or modified compounds that bind to and activate the vasopressin receptors, thereby mimicking or amplifying the biological effects of the natural hormone, vasopressin. These agents typically fall into two broad categories based on the receptors they target: those that primarily activate V1 receptors, V2 receptors, or both.

- **V1 Receptor Agonists**: These compounds predominantly target the V1a receptors, which are primarily located in vascular smooth muscle cells and the central nervous system. Activation of the V1a receptor leads to vasoconstriction, an increase in systemic vascular resistance, and an elevation in blood pressure.
- **V2 Receptor Agonists**: These compounds selectively activate V2 receptors, which are located in the kidneys, particularly in the collecting ducts. V2 receptor activation enhances water reabsorption, reduces urine output, and concentrates the urine. This is the primary mechanism by which vasopressin regulates fluid balance.

Some vasopressin agonists are designed to stimulate both V1a and V2 receptors simultaneously, allowing them to manage both fluid balance and vascular tone, which is particularly useful in cases of shock or hypotension.

5.2. Mechanism of Action of Vasopressin Agonists

The mechanism of action of vasopressin agonists is closely tied to the two major physiological actions of endogenous vasopressin: water conservation and vasoconstriction. The way these actions occur at the molecular level is dependent on the type of vasopressin receptor activated by the agonist.

Activation of V2 Receptors: Water Reabsorption

Vasopressin primarily regulates water balance through its action on the kidneys. When vasopressin or a vasopressin agonist binds to the V2 receptor on the renal tubules (specifically in the collecting ducts), a cascade of intracellular signaling events occurs:

1. **G-Protein Coupled Receptor Activation**: The V2 receptor is a G-protein coupled receptor (GPCR) that, upon ligand binding, activates the associated G-protein, which in turn activates adenylate cyclase. This increases the intracellular levels of cyclic AMP (cAMP).

2. **Activation of Protein Kinase A (PKA)**: Elevated cAMP activates PKA, which then phosphorylates specific proteins within the kidney cells.

3. **Aquaporin-2 Translocation**: The key downstream effect of this signaling pathway is the translocation of aquaporin-2 (AQP2) water channels from intracellular vesicles to the apical membrane of the renal tubular cells. Aquaporin-2 channels increase the permeability of the collecting duct to water.

4. **Water Reabsorption**: The insertion of AQP2 channels allows water to be reabsorbed from the filtrate in the kidney into the bloodstream, thus concentrating the urine and reducing urine output. This mechanism is essential for preventing dehydration and maintaining normal plasma osmolality.

Vasopressin agonists that target the V2 receptor can therefore effectively treat conditions such as **central diabetes insipidus** (where there is a deficiency of endogenous vasopressin) and **nephrogenic diabetes insipidus** (where the kidneys fail to respond to vasopressin).

Activation of V1 Receptors: Vasoconstriction and Blood Pressure Regulation

In addition to its role in water balance, vasopressin also plays a crucial role in regulating blood pressure through vasoconstriction. This effect occurs when vasopressin or a vasopressin agonist binds to the V1a receptor, which is located in the smooth muscle cells of blood vessels, especially in the arterioles.

1. **G-Protein Coupled Receptor Activation**: Like the V2 receptor, the V1a receptor is also a GPCR. Binding of the agonist activates the G-protein, which subsequently activates phospholipase C (PLC).

2. **Increased Intracellular Calcium**: Activation of PLC leads to the breakdown of phosphoinositides, which generates inositol trisphosphate (IP3). IP3 induces the release of calcium ions from intracellular stores.

3. **Smooth Muscle Contraction**: The increase in intracellular calcium leads to the contraction of vascular smooth muscle cells, resulting in vasoconstriction.

4. **Increase in Systemic Vascular Resistance**: Vasoconstriction increases the resistance against blood flow, which raises systemic vascular resistance and, consequently, blood pressure. This is particularly beneficial in situations where blood pressure is low, such as in shock or vasodilatory states.

Vasopressin agonists that target the V1a receptor are often used in **shock management**, particularly in cases of **vasodilatory shock** (such as septic shock), where they can help restore blood pressure by constricting blood vessels and improving perfusion.

Dual Action: V1a and V2 Receptor Agonism

Some vasopressin agonists have dual activity, meaning they stimulate both V1a and V2 receptors. This combined action is particularly useful in managing conditions that involve both vasodilation and fluid imbalance, such as **septic shock** and **acute heart failure**. By promoting both vasoconstriction and water retention, these agonists help stabilize blood pressure while maintaining proper fluid homeostasis.

5.3. Clinical Implications of Vasopressin Agonists

The ability to selectively target the V1a and V2 receptors with vasopressin agonists has profound implications for the treatment of several clinical conditions. The following are some of the most important therapeutic uses:

Treatment of Diabetes Insipidus

Vasopressin agonists, particularly desmopressin (DDAVP), are used to treat **central diabetes insipidus**, a condition in which there is a deficiency of endogenous vasopressin. Desmopressin selectively stimulates the V2 receptors in the kidneys, promoting water reabsorption and reducing excessive urination. This leads to improved hydration status and better control of urine output in patients with central diabetes insipidus.

Shock Management

In the setting of **shock**, particularly **vasodilatory shock** (septic shock), where there is a profound decrease in vascular tone, vasopressin agonists are used to increase blood pressure. Agents like vasopressin and terlipressin target the V1a receptors in the vascular smooth muscle, causing vasoconstriction, which helps restore normal blood pressure and improves organ perfusion.

Hyponatremia and SIADH

In **Syndrome of Inappropriate Antidiuretic Hormone (SIADH)**, where excessive secretion of vasopressin leads to water retention and dilutional hyponatremia, vasopressin receptor antagonists (vaptans) are sometimes used to block the action of vasopressin. However, vasopressin agonists may be used in other conditions where the goal is to increase water retention without causing excessive diuresis.

5.4. Challenges and Limitations

Despite their benefits, the use of vasopressin agonists is not without challenges:

- **Dose-Dependent Side Effects**: At high doses, vasopressin agonists can cause excessive vasoconstriction, leading to ischemia and tissue damage. In addition, excessive water retention may lead to **hyponatremia** and other complications.

- **Receptor Selectivity**: Not all vasopressin agonists exhibit the same receptor selectivity. Some may affect both V1a and V2 receptors, which may be beneficial in certain conditions but not ideal in others. Understanding receptor selectivity is critical in optimizing therapy.

- **Risk in Patients with Cardiovascular Disease**: In patients with pre-existing cardiovascular conditions, vasopressin agonists must be used cautiously to avoid exacerbating hypertension or causing unwanted vasoconstriction in sensitive organs.

5.5. Conclusion

Vasopressin agonists play a vital role in treating a variety of clinical conditions by mimicking the effects of endogenous vasopressin. Their mechanism of action is based on the activation of V1a and V2 receptors, leading to water retention and vasoconstriction. By understanding the underlying molecular mechanisms and the clinical implications of these agents, healthcare providers can use vasopressin agonists more effectively to manage fluid balance, blood pressure, and related disorders. The next chapter will further explore the clinical applications of vasopressin agonists, particularly in critical care settings, and highlight their role in managing severe conditions such as shock and cardiovascular disorders.

Chapter 6: Clinical Applications of Vasopressin Agonists in Critical Care

Vasopressin agonists have become invaluable tools in critical care medicine, offering a unique therapeutic advantage in the management of various life-threatening conditions. From shock to diabetes insipidus and cardiovascular instability, these agents can play a central role in stabilizing patients and improving outcomes. This chapter explores the clinical applications of vasopressin agonists in critical care, detailing how they are used to address critical conditions, the challenges associated with their use, and the evolving evidence supporting their inclusion in modern therapeutic protocols.

6.1. Vasopressin Agonists in Shock Management

Shock is a critical condition that occurs when the body's circulatory system fails to provide adequate blood flow to vital organs. It can result from several underlying causes, including trauma, infection, blood loss, and myocardial infarction. In shock, particularly **vasodilatory shock** (such as **septic shock**), the vascular tone is impaired, leading to low blood pressure and compromised organ perfusion.

6.1.1. The Role of Vasopressin Agonists in Shock

Vasopressin, through its action on V1a receptors, causes vasoconstriction by increasing intracellular calcium levels in vascular smooth muscle cells. This effect is critical in managing shock, as it raises systemic vascular resistance, which in turn elevates blood pressure and improves perfusion to vital organs. Vasopressin and its agonists, like **terlipressin** and **v1 agonists**, are particularly useful in treating shock states where standard vasopressors (e.g., norepinephrine or dopamine) are not sufficient or where their use is associated with undesirable side effects.

1. **Septic Shock**: In septic shock, a state of profound vasodilation and hypotension occurs due to the release of inflammatory mediators, such as cytokines, in response to infection. Traditional vasopressors can be less effective in this context, and vasopressin has been shown to improve blood pressure and reduce the need for additional vasopressors. The **VASST (Vasopressin and Septic Shock Trial)** demonstrated that vasopressin significantly reduced mortality rates when used as a first-line agent compared to norepinephrine in septic shock patients.

2. **Cirrhosis and Hepatorenal Syndrome**: In cirrhosis, particularly when complicated by **hepatorenal syndrome**, vasopressin agonists, especially terlipressin, have been used to improve renal function by increasing renal perfusion. Terlipressin's ability to constrict splanchnic vessels (while maintaining renal blood flow) helps reverse the hyperdynamic circulation seen in cirrhotic patients.

3. **Neurogenic Shock**: Neurogenic shock, resulting from spinal cord injury, is characterized by a loss of sympathetic tone and low blood pressure. Vasopressin agonists can help restore vascular tone and improve blood pressure, complementing the treatment with other vasopressors.

6.1.2. Choosing the Right Vasopressin Agonist

In clinical practice, the selection of a vasopressin agonist depends on the underlying cause of shock. **Vasopressin** is typically used in patients with septic shock who have not responded adequately to catecholamines, whereas **terlipressin** is more commonly used in cirrhosis with hepatorenal syndrome due to its longer half-life and specific action on splanchnic circulation.

The administration of vasopressin in shock is usually initiated at low doses, with careful titration based on hemodynamic parameters. Vasopressin should be used cautiously in patients with pre-existing coronary artery disease or those prone to ischemic complications due to its potent vasoconstrictive properties.

6.2. Vasopressin Agonists in Cardiovascular Disorders

Vasopressin agonists have a critical role in managing various cardiovascular disorders, especially in cases where regulation of vascular tone is impaired.

6.2.1. Vasopressin in Heart Failure

In heart failure, particularly in cases of **acute decompensated heart failure** (ADHF), patients often exhibit low systemic vascular resistance, leading to impaired organ perfusion. Vasopressin agonists can help restore vascular tone and improve perfusion without exacerbating the heart's workload. Vasopressin's ability to regulate fluid balance through its action on the kidneys (via V2 receptor activation) may also be beneficial in managing fluid retention, which is a hallmark of heart failure.

6.2.2. Vasopressin in Myocardial Infarction

In **myocardial infarction** (MI), especially when complicated by hypotension, vasopressin agonists may serve as adjuncts to catecholamines. Studies have shown that vasopressin can improve hemodynamic stability in the post-infarction phase by improving coronary perfusion and reducing systemic inflammation. However, caution is needed in MI patients, as vasoconstriction could exacerbate myocardial ischemia if not carefully monitored.

6.2.3. Pulmonary Arterial Hypertension (PAH)

Although not a standard treatment, vasopressin agonists have been explored in managing **pulmonary arterial hypertension (PAH)**. In this condition, there is excessive vasoconstriction in the pulmonary circulation, leading to increased pulmonary artery pressure. Vasopressin's effects on smooth muscle tone in the pulmonary vasculature may offer therapeutic potential, though more research is needed to establish efficacy and safety in this context.

6.3. Vasopressin Agonists in Diabetes Insipidus: Treatment and Challenges

Diabetes insipidus (DI) is a disorder characterized by the excretion of large volumes of dilute urine and an inability to concentrate urine due to defects in vasopressin production or action. There are two major forms of DI: **central diabetes insipidus (CDI)**, resulting from a deficiency of vasopressin secretion from the pituitary, and **nephrogenic diabetes insipidus (NDI)**, where the kidneys fail to respond to vasopressin.

6.3.1. Treatment of Central Diabetes Insipidus (CDI)

In CDI, vasopressin agonists, particularly **desmopressin** (a synthetic analogue of vasopressin), are the first-line treatment. Desmopressin selectively stimulates the V2 receptors in the kidneys, promoting water reabsorption and reducing urine output. It is typically administered intranasally or orally, depending on the severity of symptoms.

While desmopressin is highly effective in treating CDI, its use requires monitoring for potential overhydration or **hyponatremia**, especially in vulnerable populations such as the elderly or those with renal dysfunction.

6.3.2. Treatment of Nephrogenic Diabetes Insipidus (NDI)

In NDI, the kidneys are unresponsive to vasopressin, so the use of vasopressin agonists is ineffective. The treatment of NDI often involves the use of **thiazide diuretics** to reduce urine volume, **indomethacin** (a nonsteroidal anti-inflammatory drug), and adjustments in fluid intake. Although desmopressin is not typically effective in NDI, recent research has explored the use of alternative therapies, such as **demeclocycline**, to improve renal responsiveness to vasopressin.

6.4. Vasopressin Agonists in Fluid Management and Renal Function

One of the key therapeutic applications of vasopressin agonists is in managing fluid balance, particularly in conditions such as **acute kidney injury (AKI)**, **nephrotic syndrome**, and **renal pathologies**. Vasopressin agonists act on the V2 receptors in the kidneys, promoting water reabsorption and improving fluid retention without excessive sodium retention.

In critically ill patients, especially those with AKI or patients undergoing intensive care, vasopressin agonists can support fluid balance and prevent dehydration or hyponatremia. Careful monitoring of serum sodium and renal function is essential during therapy, as prolonged use of vasopressin agonists can contribute to electrolyte imbalances.

6.5. Conclusion

Vasopressin agonists have a wide range of clinical applications in critical care, from shock management and cardiovascular support to fluid balance regulation in endocrine and renal disorders. Their mechanisms of action—via vasoconstriction and water reabsorption—allow for targeted interventions that can improve hemodynamic stability, optimize fluid status, and enhance organ perfusion. However, their use must be carefully monitored to avoid complications such as **hyponatremia**, **vasoconstriction-induced ischemia**, and **overhydration**. In the next chapter, we will explore the application of vasopressin agonists in specific conditions like shock and sepsis, with an emphasis on clinical decision-making and individual patient needs.

Chapter 7: Vasopressin Agonists in Cardiovascular Disorders

Vasopressin agonists have emerged as essential therapeutic agents in managing various cardiovascular disorders. From shock states to heart failure and hypertension, vasopressin agonists can modulate vascular tone and fluid balance, thus playing a pivotal role in stabilizing patients and improving clinical outcomes. This chapter delves into the applications of vasopressin agonists in cardiovascular diseases, exploring their mechanisms of action, therapeutic benefits, potential risks, and the clinical decision-making involved in their use.

7.1. Vasopressin Agonists in Vasodilatory Shock

One of the primary applications of vasopressin agonists in cardiovascular disorders is the management of **vasodilatory shock**, which is often seen in conditions such as **septic shock, neurogenic shock**, and **anaphylactic shock**. In these conditions, the blood vessels become abnormally dilated, leading to a significant decrease in blood pressure and impaired tissue perfusion.

7.1.1. Mechanism in Vasodilatory Shock

Vasopressin agonists, such as **vasopressin** and **terlipressin**, are effective in increasing systemic vascular resistance. These agents primarily exert their effects through the **V1a receptors**, which are found in vascular smooth muscle. By binding to these receptors, vasopressin agonists increase intracellular calcium concentrations, triggering vasoconstriction and elevating blood pressure.

In **septic shock**, where systemic inflammation causes a loss of vascular tone, vasopressin's vasoconstrictive effects can significantly improve blood pressure and reduce the requirement for catecholamines like norepinephrine. Studies, including the **VASST (Vasopressin and Septic Shock Trial)**, have shown that vasopressin not only provides equivalent efficacy to norepinephrine in treating septic shock but also has the potential to reduce the overall mortality rate in certain patients.

7.1.2. Benefits Over Traditional Vasopressors

Unlike catecholamines (e.g., norepinephrine), which are predominantly sympathetic agonists, vasopressin agonists have the advantage of acting via a different pathway— modulating vascular smooth muscle via the V1a receptor. This mechanism provides a more direct and potent vasoconstrictive effect without significantly increasing myocardial oxygen demand, which can be a limiting factor in catecholamine use, particularly in critically ill patients with myocardial dysfunction.

In septic shock, where there is often a significant depletion of circulating vasopressin, adding exogenous vasopressin can help restore vascular tone without causing excessive adrenergic stimulation, making it a favorable option in this context.

7.2. Vasopressin Agonists in Heart Failure

Heart failure (HF) is another major cardiovascular disorder where vasopressin agonists may be useful. In heart failure, there is typically both reduced cardiac output and fluid retention, contributing to symptoms such as pulmonary edema, peripheral edema, and fatigue. Vasopressin agonists can help regulate fluid balance and vascular tone, which are crucial in managing these symptoms.

7.2.1. Acute Decompensated Heart Failure (ADHF)

In **acute decompensated heart failure (ADHF)**, the inability of the heart to pump blood efficiently often results in **hypoperfusion** (reduced blood flow to tissues) and **fluid overload** (due to kidney dysfunction). Vasopressin agonists, such as **desmopressin** and **terlipressin**, can be beneficial in these patients by promoting water retention in the kidneys and reducing peripheral vascular resistance.

1. **Vasopressin Agonists for Fluid Balance**: The kidney's response to vasopressin agonists via the **V2 receptors** promotes water reabsorption without causing sodium retention, which is critical in managing the fluid overload characteristic of heart failure.

2. **Vasoconstriction to Improve Perfusion**: By stimulating V1a receptors, vasopressin agonists can help increase systemic vascular resistance (SVR), improving perfusion pressure and thereby supporting organ function.

7.2.2. Potential Risks in Heart Failure

Despite the benefits, there are also risks associated with using vasopressin agonists in heart failure patients. The **V1a receptor-mediated vasoconstriction** can increase afterload (the resistance the heart must overcome to pump blood), potentially exacerbating **left ventricular (LV) dysfunction** in certain patients. This is particularly a concern in patients with **diastolic dysfunction** or those with underlying coronary artery disease.

Thus, vasopressin agonists should be used cautiously in heart failure patients, particularly those with **reduced ejection fraction**, and titrated carefully to avoid exacerbating cardiac workload.

7.3. Vasopressin Agonists in Hypertension

Although not a primary treatment, vasopressin agonists have been investigated for their role in managing **hypertension**, especially **resistant hypertension**, where blood pressure remains elevated despite the use of multiple antihypertensive agents.

7.3.1. Mechanism in Hypertension

In patients with **hypertension**, particularly **primary hyperaldosteronism** or **vascular hypertension**, vasopressin's ability to increase vascular tone can complement existing therapies. However, the vasoconstrictor effects of vasopressin must be carefully managed, as excessive vasoconstriction could lead to **end-organ damage**, including renal and cardiac complications.

While vasopressin agonists are not typically used as first-line agents for hypertension, they may be considered in patients who have failed other treatment options or in conditions where low systemic vascular resistance is contributing to poor perfusion despite adequate blood pressure.

7.4. Vasopressin Agonists in Pulmonary Arterial Hypertension (PAH)

Pulmonary arterial hypertension (PAH) is a rare but severe condition characterized by elevated blood pressure in the pulmonary arteries. This results in right heart failure due to the increased afterload on the right ventricle. Current treatments for PAH primarily include **endothelin receptor antagonists, phosphodiesterase inhibitors**, and **prostacyclin analogs**.

7.4.1. Role of Vasopressin Agonists in PAH

Although not a primary treatment for PAH, vasopressin agonists have been explored for their potential to improve pulmonary hemodynamics. In particular, vasopressin's ability to induce vasoconstriction could help mitigate the excessive vasodilation that characterizes the pulmonary circulation in PAH, potentially improving right ventricular function.

While promising in animal models and early studies, the clinical use of vasopressin agonists in PAH remains limited, and further research is required to assess their efficacy and safety in this context.

7.5. Vasopressin Agonists in Myocardial Infarction

The use of vasopressin agonists in **myocardial infarction (MI)** is an area of ongoing research. Vasopressin's effects on **vascular tone** and **renal function** can potentially improve patient outcomes in the acute setting of MI.

7.5.1. Benefits in Myocardial Infarction

In the acute phase of MI, maintaining adequate **blood pressure** and **organ perfusion** is critical. Vasopressin can help support blood pressure by inducing vasoconstriction in peripheral blood vessels, which may enhance coronary perfusion, particularly in patients who are hemodynamically unstable.

7.5.2. Potential Challenges

However, vasopressin's vasoconstrictor effects also pose a risk, particularly in patients with **pre-existing coronary artery disease**. Increased vascular tone may result in coronary vasoconstriction, which could exacerbate myocardial ischemia in patients already at risk for poor blood flow to the heart.

7.6. Conclusion

Vasopressin agonists are powerful agents with diverse applications in the management of cardiovascular disorders. They are particularly useful in treating shock states, heart failure, and pulmonary arterial hypertension, offering mechanisms of action that are complementary to existing therapies. However, their use requires careful consideration of the patient's underlying cardiovascular status and potential risks, including increased afterload and exacerbation of coronary ischemia. In the next chapter, we will explore the use of vasopressin agonists in diabetes insipidus, their role in fluid management, and the unique challenges associated with their application in endocrine disorders.

Chapter 8: Vasopressin Agonists in Diabetes Insipidus: Treatment and Challenges

Diabetes Insipidus (DI) is a rare but significant endocrine disorder characterized by the inability to concentrate urine, leading to excessive thirst (polydipsia) and the excretion of large volumes of dilute urine (polyuria). This condition is primarily associated with a deficiency or insensitivity to the hormone **vasopressin**, which plays a crucial role in regulating the body's water balance. Vasopressin agonists are vital tools in the management of DI, and their application can significantly improve the quality of life for affected patients. However, their use presents unique treatment challenges that require careful clinical judgment. This chapter explores the role of vasopressin agonists in the treatment of DI, focusing on their mechanisms of action, clinical applications, challenges in management, and emerging innovations.

8.1. Overview of Diabetes Insipidus

Diabetes Insipidus is divided into two main types based on the underlying pathophysiology:

1. **Central Diabetes Insipidus (CDI)**: This form occurs due to a deficiency or absent production of vasopressin, which is typically caused by damage to the hypothalamus or pituitary gland. The lack of vasopressin leads to an inability of the kidneys to concentrate urine, resulting in polyuria and dehydration.

2. **Nephrogenic Diabetes Insipidus (NDI)**: This form is caused by a resistance to vasopressin at the level of the kidneys. Although vasopressin is produced in normal amounts, the kidneys cannot respond to it appropriately, resulting in the same clinical manifestations of polyuria and polydipsia.

Both types of DI are marked by an impaired ability to conserve water, leading to excessive urination and thirst. Treatment strategies differ depending on the etiology of the condition.

8.2. Mechanism of Action of Vasopressin Agonists in DI

The primary role of vasopressin in the body is to regulate water homeostasis by increasing water reabsorption in the kidneys. Vasopressin agonists mimic the action of endogenous vasopressin, either by stimulating the **V2 receptors** in the kidney or by substituting for the missing or ineffective endogenous hormone.

8.2.1. Vasopressin Agonists in Central Diabetes Insipidus

In **central DI**, where there is a deficiency of vasopressin production, the administration of vasopressin agonists, such as **desmopressin** (DDAVP), provides an exogenous source of the hormone. Desmopressin is a synthetic analog of vasopressin with a much longer half-life and reduced pressor activity, making it ideal for use in DI treatment.

Upon administration, desmopressin binds to the **V2 receptors** on the distal nephron and collecting ducts in the kidneys, increasing the permeability of the renal tubules to water. This facilitates water reabsorption from the urine, resulting in a decrease in urine volume and concentration of urine.

8.2.2. Vasopressin Agonists in Nephrogenic Diabetes Insipidus

In **nephrogenic DI**, the kidneys fail to respond to vasopressin despite its normal secretion. In this case, vasopressin agonists such as desmopressin are generally ineffective because the renal tubules lack the functional receptors to bind the hormone. Instead, the treatment of nephrogenic DI typically focuses on addressing the underlying cause (e.g., electrolyte imbalances, medication side effects) or using alternative medications such as thiazide diuretics to reduce urine output.

In some cases, vasopressin analogs may still provide partial benefits by stimulating a minimal response, but they do not offer the same level of efficacy seen in CDI patients.

8.3. Clinical Applications of Vasopressin Agonists in DI Treatment

8.3.1. Desmopressin in Central Diabetes Insipidus

Desmopressin is the cornerstone of treatment for **central diabetes insipidus**. Available in various forms—intranasal, oral, and injectable—it allows for flexible dosing depending on the patient's needs and the severity of the condition.

- **Intranasal Desmopressin**: This is the most common and convenient form of administration. It is absorbed quickly through the nasal mucosa, with a typical onset of action within 30 minutes. The dosage can be adjusted based on the individual's response, with the goal of reducing urine output and preventing dehydration without causing water retention.

- **Oral Desmopressin**: Oral formulations are also available, though they are less commonly used than nasal sprays due to variable absorption. Oral desmopressin has a slower onset of action compared to intranasal administration, but it may be more suitable for patients who prefer oral medication.

- **Injectable Desmopressin**: For patients who are critically ill or require precise control of fluid balance, injectable desmopressin provides an alternative with rapid onset of action. However, it is typically reserved for hospital settings or severe cases of DI.

8.3.2. Monitoring and Dosage Adjustments

Treatment with vasopressin agonists requires close monitoring to ensure optimal dosing. Over-replacement of vasopressin can lead to **water intoxication** or **hyponatremia**, both of which are dangerous and can result in cerebral edema, seizures, or even death. Therefore, regular monitoring of urine output, plasma sodium levels, and hydration status is critical.

Adjustments in the dosage of desmopressin are based on the patient's response, with the goal of achieving a balance between adequate fluid reabsorption and preventing water retention. Clinicians must be vigilant for signs of **overhydration**, such as swelling, weight gain, or hyponatremia, and adjust the dose as necessary.

8.4. Challenges in the Treatment of Diabetes Insipidus
8.4.1. Resistance in Nephrogenic Diabetes Insipidus

In nephrogenic DI, the kidneys' inability to respond to vasopressin presents a significant challenge. While vasopressin analogs like desmopressin are largely ineffective, treatment focuses on managing symptoms and improving renal function. Pharmacologic options include thiazide diuretics, which paradoxically reduce urine output by promoting sodium and water reabsorption in the proximal tubule.

Other therapeutic strategies for nephrogenic DI include **low-sodium diets, potassium-sparing diuretics**, and **nonsteroidal anti-inflammatory drugs (NSAIDs)**, which can help reduce renal blood flow and further decrease urine output.

8.4.2. Water Balance and Hyponatremia

Achieving optimal hydration without causing hyponatremia remains one of the biggest challenges in DI management. Desmopressin, while effective at reducing urine output, can also increase the risk of water retention, leading to a dilution of sodium levels in the blood. Monitoring sodium levels regularly and titrating doses accordingly is essential to prevent this complication.

8.4.3. Patient Compliance

Adherence to desmopressin therapy can be a challenge due to the requirement for regular administration and the potential for side effects. In patients with central DI, treatment is typically lifelong, and the burden of daily medication administration can be difficult for some individuals, especially those with other chronic health conditions.

8.4.4. Special Populations

Certain populations, such as children, the elderly, or patients with renal impairment, may have unique dosing requirements. For example, pediatric patients may need lower doses, while elderly individuals are at higher risk for hyponatremia and should be monitored more closely.

8.5. Future Directions and Innovations in DI Treatment

The treatment landscape for diabetes insipidus is evolving, with ongoing research into alternative vasopressin analogs and novel delivery systems. Newer formulations of desmopressin with extended release mechanisms or more precise dosing could improve patient adherence and outcomes.

Additionally, emerging therapies, such as **V2 receptor agonists** designed to enhance the kidneys' response to vasopressin in nephrogenic DI, offer promise for improving treatment outcomes in patients who currently have limited options.

8.6. Conclusion

Vasopressin agonists are integral to the management of **central diabetes insipidus**, providing patients with a synthetic substitute for the deficient hormone. However, their use requires careful monitoring and dosage adjustments to avoid complications such as hyponatremia. In contrast, nephrogenic DI remains a more difficult condition to treat, with vasopressin agonists offering little benefit. The future of diabetes insipidus treatment lies in innovative therapies and personalized medicine approaches that optimize the management of fluid balance while minimizing risks. As research advances, new treatment strategies for both types of DI will continue to evolve, offering hope for better outcomes and improved quality of life for patients living with this challenging condition.

Chapter 9: Vasopressin Agonists in Shock Management: A Life-Saving Intervention

Shock is a critical condition characterized by insufficient blood flow and oxygen delivery to tissues, resulting in cellular dysfunction and potential organ failure. It is a medical emergency that can be caused by various factors, including trauma, infection, blood loss, and heart failure. Shock can lead to death if not promptly recognized and treated. One of the most potent and effective interventions in managing certain types of shock is the use of vasopressin agonists. This chapter delves into the role of vasopressin agonists in shock management, examining their mechanisms, applications in clinical practice, and potential outcomes.

9.1. Overview of Shock and Its Types

Shock is a complex physiological state resulting from inadequate perfusion of tissues and organs, leading to cellular ischemia, dysfunction, and potentially irreversible damage. The major types of shock include:

1. **Hypovolemic Shock**: Caused by significant blood or fluid loss, which leads to reduced circulatory volume and decreased perfusion to vital organs.

2. **Cardiogenic Shock**: Occurs when the heart's ability to pump blood is severely impaired, often as a result of a myocardial infarction or heart failure.

3. **Distributive Shock**: Involves abnormal distribution of blood flow due to vasodilation, often seen in conditions like septic shock, anaphylaxis, or neurogenic shock.

4. **Obstructive Shock**: Results from a physical obstruction to blood flow, such as pulmonary embolism or tension pneumothorax.

The pathophysiology of shock involves a cascade of events, including systemic vasodilation, myocardial depression, endothelial injury, and dysregulated immune responses. Vasopressin agonists can play a critical role in reversing some of these processes, especially in distributive and hypovolemic shock, where vascular tone is impaired.

9.2. The Role of Vasopressin in Shock

Vasopressin is a naturally occurring peptide hormone that exerts its primary effects through two types of receptors: **V1** (which mediates vasoconstriction) and **V2** (which modulates renal water retention). In shock, vasopressin's ability to constrict blood vessels via the **V1 receptors** is crucial for improving vascular tone and maintaining blood pressure. Its **V2 receptor** activity is also important in regulating fluid balance, which can be disrupted in shock states.

9.2.1. Vasopressin's Vasoconstrictive Action

In shock, particularly in distributive shock (such as septic shock), there is a significant loss of vascular tone due to widespread vasodilation. This leads to a drop in blood pressure and inadequate perfusion of organs. Vasopressin agonists, such as **synthetic vasopressin** or **desmopressin**, can act on **V1 receptors** in vascular smooth muscle, inducing vasoconstriction, thus increasing systemic vascular resistance (SVR) and improving blood pressure. This effect helps restore blood flow to vital organs, including the brain and kidneys, and is essential in stabilizing hemodynamics during shock.

9.2.2. Vasopressin in the Context of Sepsis

Septic shock, a subset of distributive shock, is one of the most critical conditions where vasopressin agonists are used. Sepsis leads to profound vasodilation, decreased vascular responsiveness to catecholamines, and the release of inflammatory mediators that impair cardiovascular function. In this context, vasopressin has a dual role: it not only induces vasoconstriction but also helps mitigate the inflammatory response, which can further worsen shock.

In sepsis, **low-dose vasopressin** (often 0.01 to 0.03 units per minute) has been shown to significantly improve mean arterial pressure (MAP) without increasing the risk of adverse events associated with high-dose catecholamines. As such, vasopressin is considered an adjunct to traditional catecholamine therapy, like norepinephrine, particularly in patients who are refractory to standard treatments.

9.3. Clinical Applications of Vasopressin Agonists in Shock Management
9.3.1. Vasopressin in Hypovolemic Shock

Hypovolemic shock, which results from significant fluid or blood loss, leads to a decrease in circulating blood volume, causing hypotension and poor tissue perfusion. While volume resuscitation (fluid or blood transfusion) is the primary treatment for hypovolemic shock, vasopressin agonists may be used as adjunctive therapy, particularly when patients do not respond adequately to fluid resuscitation.

In hypovolemic shock, vasopressin can help restore vascular tone and improve perfusion to critical organs such as the brain and kidneys. The use of vasopressin is typically considered when standard vasopressors (e.g., norepinephrine) are insufficient or contraindicated, and when shock persists despite aggressive fluid management.

9.3.2. Vasopressin in Cardiogenic Shock

Cardiogenic shock occurs when the heart's pumping ability is impaired, resulting in inadequate blood flow to the body. This can occur in conditions like acute myocardial infarction, congestive heart failure, or end-stage cardiomyopathy. The primary treatment for cardiogenic shock is to support the heart's function, often through the use of inotropes and vasopressors.

Vasopressin has limited direct use in improving cardiac output, as its primary effect is vasoconstriction rather than enhancement of myocardial contractility. However, in cases where catecholamines are ineffective, particularly in patients with high cardiac afterload, vasopressin may be used as an adjunct to improve vascular tone and support blood pressure, allowing other therapies like inotropes to be more effective.

9.3.3. Vasopressin in Distributive Shock

In distributive shock, most commonly seen in sepsis, there is widespread vasodilation that impairs the ability of blood vessels to maintain adequate blood pressure. **Septic shock**, in particular, is characterized by a combination of bacterial infection and a systemic inflammatory response that leads to vasodilation and capillary leakage.

Vasopressin's ability to promote vasoconstriction through the **V1 receptors** on smooth muscle cells is particularly valuable in septic shock. Low-dose vasopressin is often used in combination with norepinephrine to enhance the response to catecholamines, improving perfusion and reducing the need for excessive catecholamine therapy. Vasopressin's benefits in septic shock have been confirmed in clinical trials, which suggest improved outcomes when it is used early in treatment, especially for patients who are unresponsive to norepinephrine alone.

9.4. The Impact of Vasopressin Agonists on Mortality and Organ Function in Shock

The use of vasopressin agonists in shock management has been associated with several benefits, including improved blood pressure, reduced vasopressor requirements, and better organ perfusion. However, clinical studies on the impact of vasopressin on mortality have yielded mixed results.

- **Sepsis**: Clinical trials have demonstrated that vasopressin can improve hemodynamics in septic shock patients and may reduce the need for high-dose catecholamines, which are associated with adverse effects, such as arrhythmias. However, its impact on long-term survival remains uncertain, and further studies are needed to determine its effect on mortality.

- **Renal Function**: In shock, especially septic shock, vasopressin can improve renal perfusion, thereby reducing the risk of acute kidney injury (AKI). This effect is particularly important in critically ill patients who are at risk for multiorgan failure.

9.5. Challenges and Considerations in Using Vasopressin Agonists in Shock
9.5.1. Side Effects and Risks

While vasopressin agonists are generally well tolerated, they can cause side effects, particularly when used in high doses or inappropriately. Potential side effects include:

- **Hypertension**: Excessive vasoconstriction may lead to increased systemic vascular resistance and elevated blood pressure.

- **Arrhythmias**: Vasopressin's pressor effects may precipitate arrhythmias in susceptible patients.

- **Water Retention and Hyponatremia**: High doses of vasopressin or its analogs can lead to water retention, resulting in dilutional hyponatremia, a potentially life-threatening condition.

9.5.2. Dosing and Timing

The dosing and timing of vasopressin administration require careful consideration. In sepsis, for example, low-dose vasopressin is typically started early to optimize hemodynamics. In hypovolemic shock, vasopressin may be added after adequate volume resuscitation, especially in refractory cases. The goal is to balance efficacy with the risk of side effects, especially in patients with compromised organ function.

9.6. Conclusion

Vasopressin agonists have emerged as powerful tools in the management of shock, particularly in conditions like septic shock, hypovolemic shock, and distributive shock. Their ability to increase vascular tone, reduce the need for high-dose catecholamines, and improve organ perfusion has made them indispensable in critical care settings. However, their use requires careful monitoring to avoid complications such as hyponatremia, arrhythmias, and hypertension. As research continues, further refinements in vasopressin agonist therapy, including individualized dosing strategies and combination treatments, will likely improve patient outcomes and enhance the management of shock in critically ill patients.

Chapter 10: Pharmacokinetics and Pharmacodynamics of Vasopressin Agonists

Understanding the pharmacokinetics (PK) and pharmacodynamics (PD) of vasopressin agonists is crucial for optimizing their clinical use. These two aspects provide insights into how the drug is absorbed, distributed, metabolized, and eliminated from the body, as well as how it exerts its physiological effects. This chapter explores the key principles of pharmacokinetics and pharmacodynamics as they relate to vasopressin agonists, focusing on their absorption, distribution, metabolism, and elimination, as well as their dose-response relationships, efficacy, and safety profiles.

10.1. Pharmacokinetics of Vasopressin Agonists

Pharmacokinetics refers to the movement of drugs within the body, which is typically described by four primary processes: absorption, distribution, metabolism, and excretion (ADME). In the case of vasopressin agonists, understanding these processes is essential for determining the optimal dosing regimens and ensuring therapeutic efficacy while minimizing side effects.

10.1.1. Absorption

Vasopressin agonists are peptides, and like many peptides, they have relatively poor oral bioavailability due to degradation by gastrointestinal enzymes and poor absorption through the gut lining. As a result, vasopressin agonists are typically administered parenterally—either intravenously (IV), subcutaneously (SC), or intranasally.

- **Intravenous Administration**: This is the preferred route in emergency and critical care settings, especially for drugs like **vasopressin** and **desmopressin**, as it ensures rapid delivery to the bloodstream and immediate onset of action. This bypasses the digestive system entirely, making it the most reliable route for achieving therapeutic effects.

- **Subcutaneous or Intranasal Administration**: These routes are more commonly used for outpatient management of conditions like **diabetes insipidus** or **Nocturia**, where slower absorption is acceptable. Intranasal administration of **desmopressin** is frequently used for managing **central diabetes insipidus** and **bedwetting**.

10.1.2. Distribution

Once absorbed into the bloodstream, vasopressin agonists are widely distributed throughout the body. They primarily bind to plasma proteins, though the extent of this binding can vary depending on the specific agent used. For example, **vasopressin** binds to plasma proteins at a rate of approximately 10-20%, while **desmopressin**, a synthetic analog, may have a higher affinity for protein binding.

Volume of Distribution (Vd)

10.1.3. Metabolism

Vasopressin and its analogs are metabolized primarily by enzymes in the liver and kidneys. The metabolism process includes enzymatic breakdown and clearance of the drug, which can vary based on the specific agonist used.

- **Vasopressin**: In the liver and kidneys, vasopressin is metabolized by **peptidases** into smaller fragments, which are then eliminated by the kidneys. Its half-life in the circulation is relatively short (around 10-20 minutes), which contributes to the need for continuous infusion in critically ill patients.
- **Desmopressin**: As a more stable analog of vasopressin, **desmopressin** is metabolized more slowly and has a longer half-life (around 3-4 hours). This makes it useful for managing chronic conditions where sustained action is needed, such as diabetes insipidus or certain bleeding disorders (e.g., **hemophilia A**).

10.1.4. Elimination

Vasopressin agonists are primarily eliminated by the kidneys. Given that these compounds exert their effects on kidney function and water regulation, their elimination depends on renal function. **Renal clearance** is a key factor in their half-life, and impaired kidney function can lead to prolonged drug effects and increased risk of adverse events.

- **Vasopressin**: In patients with compromised renal function, the elimination of vasopressin may be slower, which could necessitate dose adjustments. In the case of vasopressin's use in critically ill patients, monitoring renal function is essential to avoid overdosage and fluid retention.
- **Desmopressin**: This agent is often favored for its longer half-life, but it should be used with caution in patients with renal impairment, as it may increase the risk of water retention and hyponatremia.

10.2. Pharmacodynamics of Vasopressin Agonists

Pharmacodynamics refers to the relationship between the concentration of a drug at its site of action and the resulting effect on the body. For vasopressin agonists, the primary pharmacodynamic effects are mediated through two receptor types: **V1** and **V2** receptors.

10.2.1. Mechanism of Action

Vasopressin exerts its physiological effects through the activation of **V1** and **V2** **receptors**:

- **V1 Receptors**: Found in vascular smooth muscle, V1 receptors mediate vasoconstriction, which leads to increased systemic vascular resistance and blood pressure. This effect is particularly important in the management of shock, where vasopressin agonists can restore vascular tone in patients with distributive shock or hypotension.

- **V2 Receptors**: Located primarily in the renal collecting ducts, activation of V2 receptors promotes water reabsorption, which helps regulate fluid balance and prevent dehydration. This effect is particularly relevant in the treatment of diabetes insipidus and other conditions involving inappropriate diuresis.

10.2.2. Dose–Response Relationship

The pharmacodynamics of vasopressin agonists are highly dose-dependent, and the effects can vary significantly based on the dose used and the clinical condition being treated.

- **Low Doses**: At lower doses (e.g., **0.01–0.03 units/min** for vasopressin), the primary effect is vasoconstriction, improving blood pressure without significantly altering renal water reabsorption. This is particularly useful in treating septic shock, where the goal is to enhance vascular tone and stabilize hemodynamics.

- **High Doses**: At higher doses, vasopressin may also exert antidiuretic effects through V2 receptor activation, which can lead to fluid retention and hyponatremia. In some cases, prolonged high-dose use may also lead to excessive vasoconstriction, resulting in ischemia in sensitive tissues like the gut and extremities.

10.2.3. Efficacy and Therapeutic Threshold

The efficacy of vasopressin agonists depends on the target condition. For shock management, vasopressin can improve outcomes by stabilizing blood pressure, reducing vasopressor requirements, and improving organ perfusion. In cases of diabetes insipidus, desmopressin's action on V2 receptors helps to prevent excessive urination and dehydration.

The therapeutic threshold—the dose at which the drug starts to exert a clinically significant effect—varies depending on the condition being treated, the patient's response, and the specific vasopressin agonist used. Monitoring clinical markers such as blood pressure, urine output, and serum electrolytes is essential for determining the appropriate dose and avoiding adverse effects.

10.3. Safety Profiles and Side Effects

While vasopressin agonists are generally well tolerated, they are not without potential side effects. Understanding the pharmacodynamics of these drugs helps clinicians anticipate and manage these risks.

10.3.1. Common Side Effects

- **Hyponatremia**: The antidiuretic effect of vasopressin agonists can lead to water retention, which may result in dilutional hyponatremia, particularly in patients with renal impairment.

- **Hypertension**: High doses of vasopressin can lead to excessive vasoconstriction, resulting in increased blood pressure and potential organ ischemia.

- **Arrhythmias**: Vasopressin's vasoconstrictive effects can sometimes lead to arrhythmias, especially in patients with underlying cardiovascular disease.

10.3.2. Risk Factors for Adverse Effects

Certain patient populations are at higher risk for experiencing adverse effects from vasopressin agonists. These include:

- **Renal Dysfunction**: Patients with impaired renal function may experience delayed elimination of vasopressin agonists, increasing the risk of fluid retention and hyponatremia.
- **Cardiovascular Disease**: Vasopressin's vasoconstrictive effects can exacerbate conditions like coronary artery disease or heart failure, particularly at high doses.

10.4. Conclusion

The pharmacokinetics and pharmacodynamics of vasopressin agonists are fundamental to understanding their clinical applications and ensuring their safe and effective use. By leveraging their unique properties—such as their ability to regulate vascular tone, blood pressure, and fluid balance—clinicians can optimize vasopressin agonist therapy for a wide range of conditions, from shock management to endocrine disorders. However, careful dosing and monitoring are essential to avoid potential side effects and achieve the best outcomes. As new formulations and delivery methods continue to evolve, the future of vasopressin agonist therapy holds promise for even more precise and individualized treatment options.

Chapter 11: Safety Profiles and Side Effects of Vasopressin Agonists

Vasopressin agonists, due to their powerful physiological effects, have a broad range of clinical applications. They are indispensable in the treatment of shock, diabetes insipidus, bleeding disorders, and certain endocrine abnormalities. However, like any pharmacological intervention, their use is not without potential risks. This chapter explores the safety profiles and common side effects of vasopressin agonists, the mechanisms behind these adverse reactions, and strategies for minimizing harm in clinical practice.

11.1. General Overview of Safety Profiles

The safety profile of any drug is determined by its therapeutic index, which is the ratio of the drug's effective dose to its toxic dose. For vasopressin agonists, the therapeutic index is relatively narrow, meaning that the difference between an effective dose and a harmful dose can be small.

Vasopressin agonists work by interacting with **V1** and **V2** receptors, and while these interactions can be therapeutically beneficial, they also lead to side effects. For instance, activation of **V1 receptors** induces vasoconstriction, which is helpful in shock but can also lead to excessive vasoconstriction and ischemia. Activation of **V2 receptors** promotes water retention in the kidneys, which can lead to fluid overload and hyponatremia.

The safety profile of these drugs can be influenced by various factors, including:

- **Dose** and **route of administration**: Higher doses and rapid intravenous infusion increase the likelihood of side effects.
- **Patient characteristics**: Renal function, cardiovascular status, and comorbid conditions (e.g., heart failure) can modify the drug's effects.
- **Duration of therapy**: Long-term use of vasopressin agonists can lead to cumulative risks, especially with agents like desmopressin, which affect fluid balance.

11.2. Common Side Effects

Despite their clinical efficacy, vasopressin agonists are associated with several common side effects that can limit their use, especially in critically ill patients.

11.2.1. Hyponatremia (Water Intoxication)

One of the most serious potential side effects of vasopressin agonists, particularly **desmopressin**, is **hyponatremia** (low sodium levels). This occurs when the body retains too much water, diluting the sodium in the bloodstream.

- **Mechanism**: Vasopressin agonists, especially those acting on **V2 receptors**, increase water reabsorption in the renal collecting ducts. If water retention is not balanced by adequate sodium intake or excretion, it can result in dilutional hyponatremia.

- **Risk Factors**: The risk is higher in patients with **renal impairment**, those receiving high doses, or those who are concurrently taking diuretics or other medications that can affect fluid and electrolyte balance.

- **Management**: Monitoring sodium levels and fluid intake is critical when administering vasopressin agonists. In patients at risk, lower doses and slower infusion rates are advisable. If hyponatremia occurs, treatment involves carefully managing fluid status and correcting electrolyte imbalances.

11.2.2. Vasoconstriction and Ischemia

While the vasoconstrictive properties of vasopressin agonists are beneficial in restoring vascular tone during shock, excessive vasoconstriction can lead to tissue ischemia, particularly in organs with limited collateral circulation.

- **Mechanism**: Activation of **V1 receptors** induces smooth muscle contraction in blood vessels, raising systemic vascular resistance and blood pressure. However, excessive or prolonged vasoconstriction can reduce blood flow to critical organs such as the kidneys, liver, and gastrointestinal system, leading to ischemia.

- **Risk Factors**: High doses of vasopressin, especially in patients with underlying cardiovascular disease or those with pre-existing ischemia (e.g., coronary artery disease), increase the risk of ischemic complications.

- **Management**: Continuous monitoring of blood pressure and organ perfusion is necessary during treatment with vasopressin agonists. In cases of excessive vasoconstriction, dose reduction or discontinuation of the drug may be required. The use of adjunctive vasodilators can also help balance the effects.

11.2.3. Arrhythmias

Vasopressin agonists can induce arrhythmias, particularly in patients with pre-existing cardiac conditions.

- **Mechanism**: The vasoconstrictive effects of vasopressin agonists may cause an increase in afterload, which can lead to elevated heart rates and stress on the myocardium. In addition, electrolyte disturbances, particularly those associated with water retention and hyponatremia, can trigger arrhythmias.

- **Risk Factors**: Patients with **heart failure, ischemic heart disease**, or those receiving concurrent antiarrhythmic medications are at increased risk for arrhythmias.

- **Management**: Cardiac monitoring, especially for patients receiving vasopressin for shock or hypotension, is critical. If arrhythmias develop, managing electrolyte imbalances and adjusting the vasopressin dose or discontinuing the drug may be necessary.

11.2.4. Headache and Nausea

Headache and nausea are relatively common side effects associated with vasopressin agonists, particularly in patients receiving high doses or those with compromised renal function.

- **Mechanism**: These symptoms are thought to result from the water retention effects and the resulting changes in blood volume and pressure. The increased blood volume and pressure may contribute to **intracranial pressure changes**, leading to headache, while nausea may be a consequence of fluid shifts and electrolyte imbalances.
- **Risk Factors**: Patients receiving long-term therapy, or those with renal impairment, are more likely to experience these symptoms.
- **Management**: Symptom relief typically involves reducing the dose or discontinuing therapy, along with careful management of fluid and electrolyte balance.

11.3. Rare but Serious Side Effects

While less common, vasopressin agonists can also cause more severe adverse effects, which necessitate close monitoring and quick intervention.

11.3.1. Water Retention and Pulmonary Edema

Excessive water retention due to vasopressin agonists can lead to **pulmonary edema**, especially in patients with compromised cardiac or renal function. This occurs when the body retains more fluid than the heart or kidneys can handle, resulting in fluid buildup in the lungs.

- **Mechanism**: The antidiuretic action of vasopressin agonists on the kidneys can overwhelm the body's ability to excrete excess fluid, leading to increased pulmonary vascular pressure and fluid accumulation in the lungs.

- **Risk Factors**: Patients with **heart failure, renal insufficiency**, or those receiving excessive doses of vasopressin agonists are at higher risk for pulmonary edema.

- **Management**: Monitoring for signs of pulmonary edema (e.g., shortness of breath, orthopnea) is essential, and immediate treatment may involve diuretics, oxygen therapy, or reduction in vasopressin agonist doses.

11.3.2. Anaphylaxis

Although rare, allergic reactions, including **anaphylaxis**, have been reported with vasopressin agonists. These reactions can range from mild symptoms (e.g., rash, itching) to life-threatening conditions (e.g., difficulty breathing, hypotension).

- **Mechanism**: The cause of anaphylaxis is typically related to hypersensitivity to the drug or its excipients (e.g., preservatives used in injectable formulations).

- **Risk Factors**: Previous allergic reactions to similar medications, including other peptide-based drugs, may increase the risk of anaphylaxis.

- **Management**: Immediate discontinuation of the drug and administration of **epinephrine, antihistamines**, and **corticosteroids** are the cornerstones of treatment. Patients who experience an allergic reaction should not be re-exposed to the drug.

11.4. Strategies for Minimizing Risk

Given the potential risks associated with vasopressin agonists, several strategies can help minimize the likelihood of adverse effects:

- **Individualized Dosing**: Dose adjustments should be based on the patient's condition, renal function, and response to treatment. Avoiding high-dose therapy, especially in critically ill patients, can reduce the risk of side effects.

- **Close Monitoring**: Continuous monitoring of vital signs (e.g., blood pressure, heart rate), electrolytes (particularly sodium), and renal function is essential during treatment with vasopressin agonists.

- **Patient Education**: Patients receiving vasopressin agonists should be educated about the signs and symptoms of adverse effects, particularly fluid retention and electrolyte imbalances, so they can seek timely medical attention.

- **Alternating Therapies**: In some cases, using adjunctive therapies, such as **norepinephrine** or **dopamine**, alongside vasopressin agonists can help achieve the desired effects with lower doses, thus minimizing the risk of side effects.

11.5. Conclusion

Vasopressin agonists are powerful tools in modern medicine, but their use must be carefully managed to avoid potential side effects. Hyponatremia, vasoconstriction, arrhythmias, and water retention are among the most common adverse effects, but with appropriate dosing, monitoring, and patient selection, these risks can be minimized. Understanding the mechanisms behind these side effects and implementing strategies for their prevention and management is key to optimizing the therapeutic benefit of vasopressin agonists in clinical practice.

Chapter 12: Emerging Research: The Future of Vasopressin Agonists

The landscape of vasopressin agonists continues to evolve as both fundamental scientific understanding and clinical practice advance. While current therapies utilizing vasopressin analogs and agonists have revolutionized the management of critical care conditions, emerging research suggests that the full potential of these molecules has yet to be realized. This chapter delves into the latest scientific findings, innovative applications, and future directions of vasopressin agonist research, examining how these compounds may shape modern medicine.

12.1. Advancements in Vasopressin Agonist Development

Historically, vasopressin has been a critical tool for managing conditions like shock, diabetes insipidus, and hemorrhagic disorders. However, newer research is focusing on improving the specificity, potency, and safety of vasopressin agonists. Innovations in peptide engineering and receptor selectivity are at the forefront of this progress.

12.1.1. Receptor–Specific Agonists

One of the most promising areas of research is the development of **receptor-specific agonists**. Traditional vasopressin agonists such as **vasopressin** and **desmopressin** target both **V1** and **V2 receptors**, leading to broad physiological effects. However, these effects can be non-selective, resulting in unwanted side effects such as excessive vasoconstriction and water retention.

- **V1 Receptor-Selective Agonists**: Research is increasingly focused on the creation of **V1-selective agonists** that specifically target the **V1a receptor**, which mediates vasoconstriction and has applications in shock management. Such selective agonists could help improve outcomes in patients with vasodilatory shock by avoiding the off-target effects on renal and water balance.

- **V2 Receptor-Selective Agonists**: Conversely, **V2-selective agonists** hold promise in treating conditions like **diabetes insipidus** and **hyponatremia**. By focusing on the antidiuretic effects mediated through the **V2 receptors**, these agents could provide more effective treatment options with a reduced risk of fluid retention and electrolyte imbalances.

Advances in receptor-specific agonism not only promise enhanced efficacy but also minimize adverse effects, paving the way for more precise, personalized treatments.

12.1.2. Peptide Modifications and Long-Acting Formulations

Another significant advancement is the development of **long-acting vasopressin agonists**. Traditional vasopressin formulations have relatively short half-lives, which requires frequent dosing, especially in critically ill patients. Researchers are exploring peptide modifications that extend the half-life of vasopressin agonists, thus reducing the frequency of administration while maintaining therapeutic efficacy.

- **PEGylation** (the process of attaching polyethylene glycol to the drug) has shown promise in extending the duration of action for vasopressin analogs.
- **Depot formulations** are being tested to allow for sustained release of vasopressin agonists over extended periods, offering significant advantages in chronic management and reducing the risk of complications associated with fluctuating drug levels.

These innovations aim to optimize drug delivery systems and enhance patient compliance, particularly in chronic and long-term treatments.

12.2. Exploring Vasopressin Agonists in New Therapeutic Areas

As the understanding of vasopressin's role in physiology deepens, new therapeutic applications for vasopressin agonists are emerging. From neuroprotection to oncology, the versatility of vasopressin and its receptors may offer novel treatment options for several challenging medical conditions.

12.2.1. Vasopressin Agonists in Neuroprotection

Neuroprotection is an exciting area of research, as **vasopressin** and its analogs have been shown to have potential benefits in **neurodegenerative diseases** and **acute brain injury**. Preclinical studies suggest that vasopressin receptor agonists may play a role in protecting neurons from injury caused by ischemia, trauma, or neuroinflammation.

- **V1 Receptors in the Brain**: The role of **V1 receptors** in the brain is being explored in the context of stroke and traumatic brain injury. These receptors are involved in processes like vasoconstriction and the regulation of cerebral blood flow, both of which may be important in managing conditions like **ischemic stroke** or **subarachnoid hemorrhage**.

- **V2 Receptors and Cognitive Function**: There is also growing interest in the involvement of **V2 receptors** in cognitive function. Some studies suggest that vasopressin's effects on memory and learning could be leveraged in the treatment of **Alzheimer's disease** and other forms of dementia, though clinical data remains sparse.

The ability of vasopressin agonists to exert neuroprotective effects could open the door to new therapies for brain injury and neurodegeneration, areas in which current treatment options are limited.

12.2.2. Vasopressin Agonists in Oncology

Recent studies have proposed that vasopressin agonists may be useful in **oncology** for managing **tumor-induced edema** and **vascular leakage**. Tumor growth often leads to pathological blood vessel formation and permeability, contributing to swelling and discomfort. Vasopressin, through its action on the **V1 receptor**, could potentially help normalize abnormal blood vessel function in tumors.

- **Vasoconstriction and Tumor Vascularity**: By selectively constricting blood vessels, vasopressin agonists might reduce the aberrant blood flow to tumors, improving the effectiveness of chemotherapy and reducing the risk of bleeding.

- **Vasopressin and Cancer-Related Hyponatremia**: Cancer patients often develop **hyponatremia**, a condition that may be exacerbated by the secretion of vasopressin-like substances (e.g., **SIADH**). Investigating vasopressin agonists for their potential to treat this condition could offer a novel approach in cancer care.

Although these concepts remain in the early stages of investigation, the emerging potential of vasopressin in oncology is an area worth monitoring.

12.3. Personalized Medicine and Genetic Insights

As pharmacogenomics continues to advance, there is a growing interest in understanding how genetic variations affect the response to vasopressin agonists. This research could lead to **personalized dosing** strategies, optimizing the safety and efficacy of treatment for individual patients based on their genetic profile.

- **Genetic Variations in Vasopressin Receptors**: Certain genetic polymorphisms in the vasopressin receptor genes (**AVPR1A** and **AVPR2**) can influence how a patient responds to vasopressin agonists. By identifying these genetic factors, clinicians may be able to tailor vasopressin therapy to maximize benefit and minimize adverse effects.

- **Biomarkers of Response**: The identification of biomarkers that predict a patient's response to vasopressin agonists could lead to more personalized treatments, reducing the risk of ineffective therapy and adverse reactions. This could be particularly useful in critical care, where individualized treatment is crucial for optimizing patient outcomes.

The integration of genetic insights into clinical practice promises to transform how vasopressin agonists are used, allowing for treatments that are safer and more effective based on a patient's unique genetic makeup.

12.4. The Role of Artificial Intelligence and Data Analytics

As in many areas of modern medicine, **artificial intelligence (AI)** and **machine learning (ML)** are playing an increasing role in drug development, clinical decision-making, and patient monitoring. AI and ML technologies could provide new insights into the optimal use of vasopressin agonists.

- **Predictive Modeling**: AI-driven algorithms may help predict patient responses to vasopressin agonists, taking into account factors like genetics, comorbidities, and drug interactions. This could improve personalized treatment plans and minimize adverse effects.

- **Clinical Trial Optimization**: AI can also streamline the design and execution of clinical trials for new vasopressin agonists, identifying the most promising compounds and patient populations faster and more accurately than traditional methods.

The integration of AI and data analytics into vasopressin agonist research and clinical care holds great promise for accelerating advancements and improving patient outcomes.

12.5. Conclusion: A Promising Future

Emerging research into vasopressin agonists is opening new frontiers in both basic science and clinical applications. With innovations in receptor specificity, long-acting formulations, and new therapeutic indications in neuroprotection, oncology, and personalized medicine, the future of vasopressin agonists looks exceptionally promising. As the understanding of these compounds deepens and technology advances, vasopressin agonists may become integral to treating a broader spectrum of diseases and improving patient outcomes across a variety of clinical settings.

Continued research and development in this field will be key to unlocking the full potential of vasopressin agonists, transforming them from a staple of critical care to a cornerstone of precision medicine.

Chapter 13: Vasopressin Agonists in Neuroprotection: Potential and Promise

Vasopressin, a hormone traditionally associated with regulating water balance and vascular tone, is now being recognized for its emerging role in neuroprotection. The ability of vasopressin agonists to modulate brain function and protect against neurodegeneration, stroke, and traumatic brain injury opens new therapeutic avenues. This chapter explores the potential of vasopressin agonists in neuroprotection, highlighting the underlying mechanisms, current research, and clinical implications.

13.1. Vasopressin and the Brain: A Brief Overview

Vasopressin, produced primarily in the hypothalamus and released by the posterior pituitary gland, plays a critical role in maintaining homeostasis by regulating blood pressure, water retention, and circadian rhythms. While vasopressin's peripheral actions are well-documented, its central effects are less understood but increasingly appreciated.

The brain is rich in vasopressin receptors, particularly **V1a** and **V2 receptors**, which are found in areas involved in behavior, emotion, and cognition, such as the hippocampus, amygdala, and prefrontal cortex. Vasopressin's ability to influence neurotransmission, vasodilation, and neuroplasticity suggests its potential in treating a variety of neurological conditions.

13.2. Neuroprotective Mechanisms of Vasopressin Agonists

Recent studies indicate that vasopressin agonists could offer neuroprotection through multiple mechanisms, making them a promising therapeutic tool in conditions like **stroke, traumatic brain injury (TBI)**, and **neurodegenerative diseases**.

13.2.1. Regulation of Cerebral Blood Flow

One of the most important ways in which vasopressin contributes to neuroprotection is through the regulation of **cerebral blood flow (CBF)**. The **V1a receptors** in cerebral blood vessels mediate vasoconstriction, which can protect against excessive blood flow during hemorrhagic stroke. Conversely, vasopressin-induced vasodilation, mediated by **V2 receptors**, can improve cerebral perfusion in conditions like **ischemic stroke**, where blood flow is impaired.

The balance between vasoconstriction and vasodilation is critical in maintaining optimal brain function and preventing secondary injury following a stroke or brain trauma. By selectively targeting vasopressin receptors, scientists aim to tailor these responses to provide optimal protection in different types of brain injury.

13.2.2. Reducing Oxidative Stress and Inflammation

Oxidative stress and inflammation are central to the pathophysiology of many neurological disorders, including **Alzheimer's disease**, **Parkinson's disease**, and **stroke**. Research has suggested that vasopressin may play a role in modulating these damaging processes.

- **Antioxidant Effects**: Vasopressin has been shown to enhance the expression of **antioxidant enzymes**, helping to reduce **reactive oxygen species (ROS)** and protect neuronal cells from oxidative damage.
- **Anti-inflammatory Properties**: Vasopressin's role in modulating inflammatory cytokines and reducing glial activation is under investigation as a potential therapeutic benefit in neurodegenerative diseases. The **V1a receptor** in particular has been implicated in regulating inflammatory responses in the brain, suggesting that vasopressin agonists could attenuate neuroinflammation associated with conditions like **multiple sclerosis** and **amyotrophic lateral sclerosis (ALS)**.

13.2.3. Modulation of Neurotransmission and Synaptic Plasticity

Vasopressin's role in modulating neurotransmitter systems is crucial for its potential neuroprotective effects. In the hippocampus, vasopressin has been found to modulate the release of **glutamate** and **gamma-aminobutyric acid (GABA)**, both of which play a pivotal role in synaptic plasticity, learning, and memory.

- **Neurotransmitter Release**: By influencing the release and uptake of neurotransmitters, vasopressin can impact synaptic function and potentially improve memory and cognitive function in conditions such as **Alzheimer's disease** or **vascular dementia**.

- **Neuroplasticity**: Vasopressin has also been shown to promote neurogenesis and synaptic remodeling, both of which are essential for recovery after brain injury. This suggests that vasopressin agonists could support brain repair mechanisms and help prevent neurodegeneration.

13.3. Vasopressin Agonists in Stroke Management

Stroke is one of the leading causes of death and disability worldwide, with **ischemic stroke** representing the majority of cases. Current therapies, such as thrombolytics and neuroprotective agents, are limited by narrow therapeutic windows and significant side effects.

Research is increasingly focusing on the use of vasopressin agonists as adjuncts to stroke management. By targeting specific vasopressin receptors, these drugs could help restore normal cerebral blood flow, reduce ischemic damage, and protect brain tissue from secondary injury.

- **Ischemic Stroke**: In animal models, vasopressin analogs have been shown to improve outcomes following ischemic stroke by enhancing cerebral perfusion and reducing the extent of neuronal damage. The potential for vasopressin agonists to serve as **neuroprotective agents** in acute ischemic events is promising, especially in patients who are not candidates for thrombolytic therapy.
- **Hemorrhagic Stroke**: Vasopressin's role in regulating vascular tone is particularly beneficial in the context of hemorrhagic stroke, where vasoconstriction may help control bleeding and reduce the risk of further hemorrhage.

13.3.1. Clinical Trials and Future Directions

Several clinical trials are underway to assess the safety and efficacy of vasopressin agonists in stroke patients. Early-phase trials are exploring the effects of these compounds on cerebral blood flow, brain edema, and long-term neurological outcomes. As data from these trials becomes available, it may pave the way for incorporating vasopressin agonists into standard stroke management protocols.

13.4. Vasopressin Agonists in Traumatic Brain Injury (TBI)

Traumatic brain injury (TBI) is a major cause of morbidity and mortality, with millions of cases reported annually worldwide. The acute phase of TBI involves a complex cascade of physiological processes, including inflammation, blood-brain barrier disruption, and neuronal death. Vasopressin agonists hold potential as part of a therapeutic strategy to reduce the damage associated with these events.

13.4.1. Vascular and Cellular Protection

Vasopressin's ability to modulate blood vessel permeability and reduce **cerebral edema** could provide significant benefits in TBI management. By selectively targeting the vasopressin receptor subtypes, researchers hope to develop therapies that can mitigate brain swelling while improving perfusion to injured tissues.

- **V1 Receptor Activation**: Vasopressin's action on **V1a receptors** can reduce vascular permeability and prevent further leakage of fluid into the brain, potentially decreasing the severity of **cerebral edema**.
- **Neuroprotective Effects**: Additionally, vasopressin agonists could enhance neuroprotection by reducing oxidative stress and inflammation in the brain, thus limiting neuronal death and supporting recovery.

13.4.2. Preclinical and Clinical Studies

Preclinical studies in animal models of TBI have demonstrated that vasopressin analogs can reduce tissue damage and improve recovery following injury. While clinical evidence is still limited, the findings suggest that vasopressin agonists could be a valuable adjunct in the treatment of TBI, particularly in the acute and subacute phases.

13.5. Vasopressin Agonists in Neurodegenerative Diseases

Neurodegenerative diseases such as **Alzheimer's disease**, **Parkinson's disease**, and **frontotemporal dementia** represent some of the most pressing challenges in modern medicine. Vasopressin agonists are being investigated for their potential to slow disease progression and improve cognitive function.

13.5.1. Alzheimer's Disease and Vasopressin

In Alzheimer's disease, vasopressin's ability to modulate the hippocampus and improve synaptic function may help address some of the cognitive deficits associated with the disease. Research is exploring the effects of vasopressin agonists on **amyloid-beta accumulation**, **tau phosphorylation**, and neuroinflammation—all of which play a role in the progression of Alzheimer's disease.

13.5.2. Parkinson's Disease and Motor Function

In Parkinson's disease, the loss of dopaminergic neurons in the **substantia nigra** leads to motor dysfunction and neurodegeneration. Vasopressin's effects on neurotransmitter systems, particularly dopamine, are being explored as potential therapeutic mechanisms to improve motor function and reduce neurodegeneration in Parkinson's patients.

13.6. Conclusion: The Future of Vasopressin Agonists in Neuroprotection

Vasopressin agonists hold significant promise in the field of neuroprotection. Through their effects on cerebral blood flow, inflammation, oxidative stress, and synaptic function, these agents may offer novel treatment options for stroke, traumatic brain injury, and neurodegenerative diseases. As research progresses, we can expect to see vasopressin agonists integrated into therapeutic strategies aimed at preserving brain health and improving recovery after neurological injuries.

The growing body of evidence supports the need for further clinical trials and preclinical research to better understand the role of vasopressin in neuroprotection, and to ultimately harness its potential in addressing some of the most challenging neurological conditions in modern medicine.

Chapter 14: Vasopressin Agonists in Kidney Function and Renal Pathologies

The kidneys are central to regulating the body's fluid balance, electrolyte composition, and overall homeostasis. Vasopressin, with its effects on water reabsorption in the kidneys, plays a crucial role in these processes. Vasopressin agonists, through their modulation of vasopressin receptors, offer promising therapeutic potential in treating a variety of renal pathologies. This chapter explores the role of vasopressin agonists in kidney function, their applications in renal disorders, and the evolving research aimed at harnessing their benefits for patients with kidney-related conditions.

14.1. Overview of Vasopressin's Role in Renal Physiology

Vasopressin, also known as **antidiuretic hormone (ADH)**, is primarily responsible for regulating water balance in the body. It acts on the kidneys by promoting water reabsorption through **aquaporins**, particularly **aquaporin-2 (AQP2)** channels, located in the **collecting ducts** of the nephrons. This action is mediated via **V2 receptors** on the renal tubular cells. When vasopressin binds to these receptors, it triggers the insertion of aquaporin-2 channels into the cell membrane, facilitating water reabsorption and thereby reducing urine output.

In conditions of dehydration or increased plasma osmolality, vasopressin levels rise to conserve water. Conversely, when the body is hydrated, vasopressin secretion decreases, leading to increased urine output. This finely tuned balance between water retention and excretion is essential for maintaining **fluid homeostasis** and **electrolyte balance**.

14.2. Vasopressin Agonists in Renal Pathologies

The therapeutic application of vasopressin agonists extends to a range of renal disorders, particularly those associated with water balance, fluid overload, and electrolyte imbalances. These disorders include **Diabetes Insipidus (DI), acute kidney injury (AKI), chronic kidney disease (CKD)**, and **nephrogenic diabetes insipidus (NDI)**, among others.

14.2.1. Diabetes Insipidus (DI)

Diabetes Insipidus (DI) is characterized by the kidneys' inability to concentrate urine, resulting in excessive urination (polyuria) and thirst (polydipsia). There are two primary forms of DI: **central DI**, which results from insufficient vasopressin production by the hypothalamus, and **nephrogenic DI**, where the kidneys are resistant to the action of vasopressin.

- **Central DI**: In central DI, vasopressin agonists, such as **desmopressin (DDAVP)**, are the cornerstone of treatment. Desmopressin is a synthetic analog of vasopressin that selectively activates the **V2 receptors** in the kidney, thereby increasing water reabsorption in the collecting ducts. This reduces urine output and alleviates the symptoms of excessive thirst and urination. Desmopressin has become the gold standard for central DI management due to its efficacy and safety profile.

- **Nephrogenic DI**: In nephrogenic DI, the kidneys do not respond to vasopressin due to **V2 receptor dysfunction** or abnormalities in **aquaporin-2** channels. Treatment strategies focus on addressing the underlying cause, such as discontinuing nephrotoxic medications or correcting electrolyte imbalances. In some cases, vasopressin agonists can be used in combination with other pharmacological agents, such as **thiazide diuretics**, to manage urine output. However, response to vasopressin analogs in nephrogenic DI is often limited and variable.

14.2.2. Acute Kidney Injury (AKI)

Acute Kidney Injury (AKI) is characterized by a sudden decline in kidney function, leading to the retention of waste products, fluid overload, and electrolyte disturbances. Vasopressin agonists have shown potential in improving renal perfusion and enhancing urine output in AKI patients, particularly in those with **prerenal azotemia** (a condition where kidney failure is primarily due to insufficient blood flow).

- **Vasopressin Agonists in AKI**: In animal models and some clinical studies, vasopressin analogs have been shown to improve renal perfusion and reduce the severity of AKI by promoting vasoconstriction in the renal vasculature. The use of vasopressin agonists can help restore adequate glomerular filtration rates (GFR) and urine output, especially in cases where the underlying cause is volume depletion or hypotension.

- **V1 Receptor Activation**: The **V1a receptors** in the renal vasculature can mediate vasoconstriction, which may improve renal blood flow in certain circumstances. However, the use of vasopressin agonists in AKI is a subject of ongoing debate, as vasoconstriction may also exacerbate ischemic injury in the kidneys under certain conditions. The key challenge is identifying the patients who would benefit from vasopressin agonist therapy while minimizing the risks of further kidney damage.

14.2.3. Chronic Kidney Disease (CKD)

Chronic kidney disease (CKD) is a progressive loss of kidney function that affects a significant proportion of the global population. As CKD progresses, the kidneys' ability to concentrate urine becomes impaired, leading to **polyuria** and further fluid and electrolyte imbalances. In such patients, vasopressin agonists may help manage fluid retention and balance the water-electrolyte equilibrium.

Vasopressin Agonists in CKD Management

hyponatremia

fluid overload

hypernatremia

water intoxication

14.2.4. Nephrogenic Syndrome of Inappropriate Antidiuresis (NSIAD)

Nephrogenic Syndrome of Inappropriate Antidiuresis (NSIAD) is a rare disorder characterized by excessive secretion of vasopressin despite the presence of normal or low plasma osmolality. This condition leads to **fluid retention, hyponatremia**, and **edema**. Vasopressin agonists are contraindicated in NSIAD, as they can exacerbate the fluid overload and hyponatremia. The treatment approach in such cases involves managing the underlying condition and correcting fluid and electrolyte imbalances.

14.3. Future Directions in Vasopressin Agonist Therapy for Renal Disorders

As our understanding of vasopressin's role in renal pathophysiology deepens, the potential for **new vasopressin agonists** to address various kidney disorders grows. Research into the selective activation of specific vasopressin receptor subtypes, such as **V1a** and **V2 receptors**, is opening doors to more targeted therapies with fewer side effects.

14.3.1. V2 Receptor Agonists for Water Retention

One exciting area of research is the development of **V2 receptor-selective agonists**, which could be used in conditions where water retention is necessary. These agonists have the potential to enhance water reabsorption without significantly affecting blood pressure or vascular tone, making them particularly useful for managing conditions like **nephrogenic DI** or **acute kidney injury**.

14.3.2. Vasopressin Agonists in Dialysis

Another promising application of vasopressin agonists is in the management of patients undergoing **dialysis**. By enhancing renal function and increasing the efficiency of dialysis, vasopressin agonists could help reduce the need for frequent dialysis sessions and improve patient outcomes.

14.3.3. Gene Therapy Approaches

Gene therapy targeting the expression of vasopressin receptors or aquaporin-2 channels in the kidneys is another innovative area of research. This approach may provide long-term solutions for patients with genetic forms of **nephrogenic diabetes insipidus** or other inherited kidney disorders.

14.4. Conclusion: The Evolving Role of Vasopressin Agonists in Renal Medicine

Vasopressin agonists have proven to be valuable tools in managing a range of renal pathologies, particularly those involving water balance and fluid homeostasis. From **central diabetes insipidus** to **acute kidney injury** and **chronic kidney disease**, these agents offer therapeutic promise for improving kidney function and managing fluid and electrolyte imbalances.

As research progresses, the development of more selective and targeted vasopressin agonists holds the potential to revolutionize the treatment of kidney diseases, providing better outcomes for patients with renal pathologies and improving their quality of life. However, as with all medications, careful patient selection and monitoring are essential to optimize the benefits while minimizing potential risks.

In the future, as the mechanisms of vasopressin action become more precisely understood, vasopressin agonists may play an even larger role in kidney medicine, offering new hope for patients suffering from kidney diseases.

Chapter 15: Case Studies in Vasopressin Agonist Therapy

The application of vasopressin agonists in clinical practice is diverse, spanning critical care, endocrine management, shock treatment, and various renal and neurological conditions. While the underlying mechanisms and biochemical processes are well-established, real-world application is often more complex, requiring individualized treatment plans and careful monitoring. This chapter presents a series of case studies that highlight the use of vasopressin agonists across different clinical scenarios. These case studies provide valuable insights into the practical considerations, challenges, and successes of vasopressin agonist therapy.

15.1. Case Study 1: Vasopressin Agonists in the Management of Central Diabetes Insipidus

Patient Background: A 45-year-old male diagnosed with central diabetes insipidus (CDI) secondary to a pituitary adenoma presented with complaints of extreme thirst (polydipsia) and excessive urination (polyuria). His lab work revealed a serum osmolality of 295 mOsm/kg (elevated), and urine osmolality was 50 mOsm/kg (low, indicative of dilute urine). An MRI confirmed the presence of a small pituitary tumor. The patient's serum sodium levels were elevated at 148 mEq/L, and he was severely dehydrated.

Clinical Approach: Given the diagnosis of CDI, the patient was initiated on **desmopressin (DDAVP)**, a synthetic vasopressin analog. Desmopressin was chosen for its high selectivity for the **V2 receptors** in the kidneys, minimizing vasoconstrictive effects that could impact blood pressure. The starting dose of 0.1 mg intranasally was administered.

Outcome: After 24 hours of treatment, the patient experienced a significant reduction in polyuria and polydipsia. His urine output decreased from 6 liters per day to 2 liters per day, and serum sodium levels normalized within 48 hours. Urine osmolality increased to 450 mOsm/kg, indicating proper water reabsorption. The patient was discharged with instructions for continued desmopressin use, with regular follow-up to monitor electrolyte levels.

Key Takeaway: Desmopressin proved to be highly effective in managing the symptoms of central diabetes insipidus, stabilizing electrolyte levels and preventing dehydration. This case illustrates the importance of timely diagnosis and appropriate use of vasopressin agonists in managing CDI.

15.2. Case Study 2: Vasopressin Agonists in Shock Management

Patient Background: A 62-year-old female with a history of hypertension and type 2 diabetes mellitus presented to the emergency department with signs of septic shock. She was hypotensive (BP 80/50 mmHg), tachycardic (HR 120 bpm), febrile (39.2°C), and had lab results consistent with severe sepsis (elevated lactate, WBC count of 18,000/μL). Despite aggressive fluid resuscitation, her blood pressure remained low, and she was unresponsive to norepinephrine.

Clinical Approach: Given the patient's persistent hypotension, **vasopressin (Pitressin)** was initiated at a low dose (0.01 units/min IV) in conjunction with norepinephrine. This combination was chosen due to the synergistic effects of vasopressin in improving vascular tone, especially in **vasodilatory shock** such as sepsis.

Outcome: Within hours of starting vasopressin, the patient's blood pressure improved to 110/70 mmHg, and her lactate levels began to decrease, indicating improved tissue perfusion. Vasopressin was tapered down over the next 48 hours as the patient's condition stabilized. She was eventually transferred to the ICU for continued monitoring and treatment.

Key Takeaway: In this case, vasopressin played a critical role in managing septic shock and refractory hypotension. By acting on **V1a receptors** in the vasculature, vasopressin improved systemic vascular resistance and helped stabilize the patient's hemodynamics. This case highlights the importance of vasopressin in critical care settings, particularly in patients unresponsive to catecholamine therapy.

15.3. Case Study 3: Vasopressin Agonists in Nephrogenic Diabetes Insipidus

Patient Background: A 30-year-old male presented with a history of excessive urination (up to 10 liters per day) and extreme thirst. Laboratory tests revealed low urine osmolality (<100 mOsm/kg) despite hyperosmolar plasma (serum osmolality 295 mOsm/kg). The patient was diagnosed with nephrogenic diabetes insipidus (NDI) due to **lithium therapy**, which had been used for bipolar disorder. His serum sodium was elevated at 148 mEq/L, and he was also dehydrated.

Clinical Approach: NDI occurs when the kidneys are resistant to the effects of vasopressin, often due to defects in the **V2 receptors** or **aquaporin-2** channels. Given that vasopressin agonists are generally ineffective in NDI, the treatment focus was on correcting fluid balance. In addition to aggressive fluid resuscitation, the patient was started on a combination of **hydrochlorothiazide (a thiazide diuretic)** and **amiloride (a potassium-sparing diuretic)**, with a goal of reducing polyuria.

Outcome: Although desmopressin did not improve the patient's symptoms, the use of diuretics successfully reduced urine output. After a few days of fluid management and diuretic therapy, his serum sodium levels decreased to normal ranges, and the patient's urine output dropped to 4 liters per day. Lithium was discontinued, and the patient was referred for follow-up care with a nephrologist.

Key Takeaway: In nephrogenic diabetes insipidus, vasopressin agonists are typically ineffective due to the inherent resistance in the kidneys. However, supportive therapies such as thiazide diuretics can help mitigate symptoms. This case underscores the importance of identifying the underlying etiology of diabetes insipidus and tailoring the treatment approach accordingly.

15.4. Case Study 4: Vasopressin Agonists in Cardiovascular Surgery

Patient Background: A 68-year-old male undergoing **coronary artery bypass grafting (CABG)** surgery developed significant hypotension after cardiopulmonary bypass (CPB), despite receiving large volumes of fluids and conventional vasopressors such as norepinephrine. The patient's blood pressure remained 80/50 mmHg, and his cardiac output was low. Given the persistent hypotension, vasopressin was considered as an adjunct to improve vascular tone.

Clinical Approach: **Vasopressin (Pitressin)** was initiated at a low dose (0.02 units/min IV), and norepinephrine was concurrently reduced to minimize excessive vasoconstriction. The decision to use vasopressin was based on its ability to increase systemic vascular resistance, particularly in patients who are refractory to conventional vasopressors after cardiac surgery.

Outcome: The administration of vasopressin led to an immediate improvement in blood pressure, which rose to 105/70 mmHg within 30 minutes. Cardiac output also improved, and the patient's perfusion status stabilized. Vasopressin was continued for 48 hours post-surgery before being gradually tapered off as the patient's hemodynamic status normalized.

Key Takeaway: In this case, vasopressin was a valuable adjunct in the management of hypotension following cardiac surgery. Its ability to improve vascular tone and support hemodynamics in patients unresponsive to catecholamine therapy highlights its utility in postoperative care.

15.5. Conclusion: Real–World Insights and Practical Applications

These case studies demonstrate the broad range of clinical scenarios in which vasopressin agonists can be employed effectively. Whether in managing **diabetes insipidus**, treating **shock**, or stabilizing **renal function**, vasopressin agonists have proven to be versatile agents in clinical medicine. However, as with all therapies, careful patient selection and monitoring are essential to ensure optimal outcomes and minimize risks.

As clinical experience with vasopressin agonists continues to grow, healthcare providers will gain a deeper understanding of their nuances, allowing for more precise, individualized treatment strategies. These case studies underscore the importance of a comprehensive approach to treatment, considering the patient's entire clinical picture and the therapeutic potential of vasopressin agonists.

Chapter 16: Optimizing Vasopressin Agonist Dosing for Individualized Treatment

One of the key challenges in the clinical application of vasopressin agonists is optimizing dosing to achieve the desired therapeutic effect while minimizing side effects. Vasopressin agonists, whether in the form of vasopressin or synthetic analogs like desmopressin or terlipressin, exhibit potent biological actions that require careful titration. The purpose of this chapter is to explore the factors influencing vasopressin agonist dosing, the methods for tailoring doses to individual patients, and the strategies for monitoring efficacy and safety in various clinical contexts.

16.1. Key Factors Affecting Vasopressin Agonist Dosing

The appropriate dosing of vasopressin agonists is influenced by a variety of patient-specific factors, including the underlying condition being treated, the patient's physiological state, and the pharmacokinetic properties of the drug. These factors must be considered in a comprehensive treatment plan to ensure the best possible outcomes.

16.1.1. Type of Vasopressin Agonist

Different vasopressin agonists have distinct pharmacological profiles, and this impacts the starting dose and titration regimen. For example:

- **Desmopressin (DDAVP)**, a synthetic analogue of vasopressin, is primarily used to treat diabetes insipidus and has a minimal pressor effect. It is typically administered intranasally or subcutaneously at doses ranging from 0.05 to 0.2 mg/day, depending on the severity of symptoms.

- **Vasopressin (Pitressin)**, used mainly in the treatment of shock or vasodilatory conditions, has both antidiuretic and vasoconstrictive effects. Its typical starting dose for shock is 0.01–0.03 units/min IV, which can be titrated based on blood pressure response.

- **Terlipressin**, which is primarily used in cirrhosis and hepatorenal syndrome, has a longer duration of action than vasopressin and is generally given at doses of 1–2 mg IV every 4–6 hours.

16.1.2. Patient's Age and Comorbidities

Age-related changes in drug metabolism, especially in the elderly, can influence the pharmacodynamics of vasopressin agonists. Older patients may require lower doses or more frequent monitoring to avoid excessive vasoconstriction, particularly in the context of cardiovascular diseases such as hypertension or coronary artery disease.

Comorbid conditions such as **renal impairment**, **cardiovascular disease**, or **endocrine disorders** may also affect how the body responds to vasopressin agonists. For example, patients with renal dysfunction may require lower doses of desmopressin, as impaired renal clearance can lead to fluid retention and hyponatremia.

16.1.3. Electrolyte Imbalance and Hydration Status

Patients with imbalances in serum sodium or potassium levels require special attention when dosing vasopressin agonists. **Hyponatremia**, a common complication with vasopressin analogs, especially desmopressin, may require dose reduction or alternative therapies. Similarly, patients who are dehydrated may need higher initial doses to restore fluid balance, whereas those who are fluid overloaded may need to have vasopressin agonists carefully titrated to avoid worsening of fluid retention and edema.

16.1.4. Type of Shock or Critical Illness

Vasopressin is commonly used in managing shock states, particularly **septic shock** and **vasodilatory shock**. In these conditions, blood pressure response to standard vasopressors like norepinephrine may be insufficient, necessitating the addition of vasopressin. The initial dose of vasopressin in shock can range from 0.01 to 0.03 units/min IV, and it is typically titrated upwards to maintain a mean arterial pressure (MAP) above 65 mmHg. However, in conditions like **cardiogenic shock** or **hypovolemic shock**, vasopressin may not be as effective and can exacerbate underlying heart failure.

16.2. Strategies for Dosing Optimization
16.2.1. Starting Dose and Titration

The goal in initiating vasopressin therapy is to start with the lowest effective dose and titrate upwards based on the patient's response. This approach minimizes the risk of side effects, such as excessive vasoconstriction or fluid retention.

- **For vasopressin in shock management**, a typical starting dose of 0.01–0.03 units/min IV is used. In patients who do not respond to initial dosing, the dose may be increased in increments of 0.01–0.02 units/min, with close monitoring of blood pressure, heart rate, and fluid status.
- **For desmopressin**, therapy usually begins at a low dose (0.1 mg) and can be titrated upwards in cases of persistent symptoms, with careful monitoring of serum sodium levels to avoid hyponatremia.

16.2.2. Continuous Infusion vs. Bolus Dosing

Vasopressin and its analogs can be administered via continuous infusion or as intermittent bolus doses, depending on the clinical scenario:

- **Continuous infusion** is preferred in cases of severe shock or diabetes insipidus, as it allows for precise control over the vasopressor or antidiuretic effect. Infusion rates are adjusted based on hemodynamic and electrolyte parameters.
- **Bolus dosing** is typically used in certain surgical or critical care settings where immediate action is required, such as post-cardiac surgery or during acute hemorrhage.

16.2.3. Monitoring and Adjustments

Monitoring the effects of vasopressin agonists is essential for ensuring that dosing remains appropriate for each patient. Key parameters to monitor include:

- **Blood pressure and heart rate**: In shock management, it is important to regularly check the patient's MAP and adjust the dose of vasopressin accordingly. Too high a dose can lead to excessive vasoconstriction, increasing the risk of ischemia.

- **Electrolyte levels**: Close monitoring of sodium and potassium levels is essential, especially when using desmopressin or in patients with kidney dysfunction. Low sodium levels (hyponatremia) can occur, especially with overdosage of desmopressin.

- **Fluid status**: Vasopressin can cause fluid retention, and in patients with renal impairment or heart failure, this can result in **pulmonary edema** or **cardiogenic shock**. Fluid balance should be monitored closely, and diuretics may be required in certain cases.

16.3. Individualized Treatment Plans

The optimal dose of vasopressin agonists is not a one-size-fits-all approach. Rather, dosing should be individualized based on the patient's specific clinical situation. The following factors should be considered when developing an individualized treatment plan:

- **Severity of the condition**: For conditions like shock or severe diabetes insipidus, higher doses may be required initially, with gradual adjustments as the patient's condition improves. For less severe cases, lower doses may be sufficient.

- **Response to therapy**: Dose adjustments should be based on continuous monitoring of clinical endpoints such as blood pressure, urine output, and serum sodium levels. Regular assessments allow for timely modifications to the treatment regimen.

- **Patient characteristics**: Comorbidities such as renal insufficiency, liver disease, or cardiovascular disorders necessitate careful consideration when determining the appropriate dose. Patients with these conditions may require reduced doses or more frequent monitoring.

16.4. Challenges and Future Directions

Despite the advancements in vasopressin agonist therapy, challenges remain in optimizing dosing strategies, particularly in critically ill patients. The variability in patient responses, potential for drug interactions, and the risk of side effects like **hyponatremia, hypertension**, and **cardiovascular complications** present ongoing concerns.

Emerging technologies, such as **biomarker-driven dosing protocols** and **smart infusion pumps**, offer the potential for more precise control over vasopressin administration. The integration of **real-time monitoring** of hemodynamic parameters and electrolyte levels could improve dosing accuracy, minimize risks, and enhance patient outcomes.

16.5. Conclusion

Optimizing vasopressin agonist dosing is a crucial element in maximizing therapeutic outcomes while minimizing adverse effects. By considering patient-specific factors, adjusting doses based on clinical responses, and monitoring key parameters, healthcare providers can better tailor vasopressin therapy to the individual needs of their patients. The future of vasopressin agonist therapy will likely see even more refined dosing protocols and technologies that enhance both efficacy and safety, ensuring that these powerful agents continue to provide life-saving benefits in a variety of clinical settings.

Chapter 17: The Role of Vasopressin Agonists in Anesthesia and Surgery

The use of vasopressin agonists in the perioperative setting represents a critical and emerging area in anesthesia and surgery. These agents, known for their vasoconstrictive, antidiuretic, and pressor effects, have the potential to significantly impact outcomes during both major and minor surgical procedures. This chapter explores the role of vasopressin agonists in the context of anesthesia, surgery, and perioperative care, focusing on their mechanisms of action, clinical applications, and considerations for optimizing patient care.

17.1. Physiological Role of Vasopressin During Surgery

During surgery, patients undergo various physiological stresses such as blood loss, fluid shifts, and changes in cardiac output. Vasopressin plays a central role in maintaining hemodynamic stability by modulating vascular tone, regulating water balance, and influencing blood pressure. As an endogenous peptide, vasopressin's secretion increases in response to stressors like hypovolemia, pain, and changes in blood pressure, making it an important factor in perioperative care.

In this context, vasopressin agonists can be strategically employed to support blood pressure, mitigate the effects of fluid loss, and balance electrolytes in patients undergoing invasive procedures. The pharmacological use of vasopressin agonists, such as vasopressin and its analogs (e.g., desmopressin and terlipressin), can help optimize perioperative outcomes by:

- **Maintaining vascular tone**: By constricting blood vessels, vasopressin agonists help prevent hypotension during and after surgery.

- **Regulating fluid balance**: Vasopressin has potent antidiuretic effects, which can aid in preventing dehydration and managing fluid retention in the surgical patient.

- **Enhancing tissue perfusion**: In certain shock states or during massive blood loss, vasopressin agonists can improve perfusion to vital organs, such as the brain, kidneys, and heart.

17.2. Clinical Applications of Vasopressin Agonists in Anesthesia and Surgery

Vasopressin agonists are increasingly being recognized for their utility in various surgical and anesthetic settings, including **shock management**, **cardiovascular support**, and **fluid balance** regulation. Here, we discuss the most common applications of vasopressin agonists in these settings:

17.2.1. Management of Hypotension in Major Surgery

Hypotension is a common complication in major surgeries, especially those that involve significant blood loss, fluid shifts, or extensive manipulation of large blood vessels. In patients with **hypovolemic shock**, **septic shock**, or **cardiogenic shock**, traditional vasopressors such as norepinephrine or epinephrine may be insufficient, requiring the addition of vasopressin or its analogs.

- **Vasopressin**: This is often used as an adjunct in managing **septic shock** or **vasodilatory shock**, where blood pressure remains low despite the administration of other vasopressors. It acts via the **V1 receptors** in vascular smooth muscle to induce vasoconstriction, which helps increase systemic vascular resistance (SVR) and elevate blood pressure.

- **Desmopressin (DDAVP)**: Although more commonly used in the treatment of bleeding disorders, desmopressin has also been explored for its role in **reducing bleeding** during surgery, particularly in patients with **von Willebrand disease** or **mild hemophilia**. By enhancing the release of von Willebrand factor, desmopressin can help improve platelet adhesion and clot formation.

17.2.2. Perioperative Blood Loss and Coagulation Support

Excessive blood loss during surgery can lead to a series of complications, including hypovolemic shock, coagulopathy, and multiorgan dysfunction. Vasopressin analogs can be used as adjuncts to help control blood loss and support coagulation:

- **Desmopressin**: In addition to its role in promoting hemostasis, desmopressin is particularly valuable in patients undergoing **cardiac surgery**, **orthopedic surgery**, or any procedure where bleeding risks are high. Its ability to increase platelet function makes it an essential tool in managing bleeding diatheses in surgical patients.

- **Terlipressin**: This vasopressin analog is often used in **liver transplant** patients, where significant blood loss can result in a drop in blood pressure and circulatory instability. By constricting splanchnic vessels, terlipressin can help increase effective circulating blood volume, thereby improving hemodynamic stability and organ perfusion.

17.2.3. Fluid Balance and Antidiuresis

Fluid balance is crucial in the perioperative setting to avoid both dehydration and fluid overload. In patients undergoing surgery, fluid shifts often occur due to blood loss, interstitial fluid accumulation, and changes in perfusion. Vasopressin's antidiuretic effects can be beneficial in managing these shifts:

- **Desmopressin** is often used in the management of **diabetes insipidus** during surgery, especially in patients with hypothalamic or pituitary dysfunction who are at risk of excessive fluid loss. By promoting water reabsorption in the kidneys, desmopressin helps prevent dehydration and maintains electrolyte balance.

- In patients at risk of **postoperative fluid retention** (e.g., those with cardiac or renal disease), **vasopressin antagonists** or adjustments to vasopressin agonist dosing can be considered to avoid fluid overload.

17.2.4. Vasopressin Agonists in Anesthetic Agents

The perioperative administration of vasopressin agonists can also interact with **anesthetic agents**, enhancing or modifying their effects. For example:

- **Opioids**: Opioids, commonly used for pain management in surgery, can trigger the release of endogenous vasopressin. In patients who receive high doses of opioids, there may be an enhanced risk of **hyponatremia**, particularly in individuals with existing kidney dysfunction. The use of exogenous vasopressin in these settings can help prevent this imbalance.

- **General Anesthesia**: In patients undergoing general anesthesia, especially those with **shock states** or **hypotension**, vasopressin or its analogs may be used to maintain vascular tone and ensure that blood pressure remains stable during induction and maintenance of anesthesia.

17.3. Challenges and Considerations in Anesthesia and Surgery

While the use of vasopressin agonists in anesthesia and surgery holds significant promise, there are several challenges and considerations that clinicians must address:

17.3.1. Risk of Overuse and Adverse Effects

Vasopressin agonists have powerful effects on the vascular system, and their use requires careful monitoring to avoid complications such as **excessive vasoconstriction, hyponatremia, arrhythmias**, or **cardiac ischemia**. The risk of **renal vasoconstriction** in patients with pre-existing kidney disease is a particular concern, as vasopressin analogs can decrease renal perfusion and exacerbate kidney dysfunction.

17.3.2. Monitoring and Titration

Continuous monitoring is essential when using vasopressin agonists during anesthesia and surgery. Parameters such as **blood pressure, urine output, serum sodium levels,** and **electrolyte balance** should be frequently assessed to ensure the appropriate response to therapy. Dosing should be titrated to effect, with the goal of maintaining hemodynamic stability while minimizing the risk of side effects.

17.4. Future Directions in Vasopressin Agonist Use in Surgery

As surgical techniques evolve and the understanding of vasopressin's role in perioperative care deepens, new strategies for using vasopressin agonists will likely emerge. **Personalized dosing protocols**, guided by biomarkers and advanced monitoring systems, could allow for more precise control of vasopressin therapy, improving patient outcomes and minimizing risks.

Additionally, **novel vasopressin receptor antagonists** may offer new opportunities for balancing vasoconstriction and fluid retention in specific patient populations, particularly those with complex comorbidities.

17.5. Conclusion

Vasopressin agonists are powerful tools in the management of surgical patients, offering significant benefits in maintaining hemodynamic stability, regulating fluid balance, and managing coagulopathy. By understanding the physiological actions of these agents and optimizing their use in various perioperative contexts, anesthesiologists and surgeons can enhance patient outcomes, improve recovery times, and reduce complications. The continued exploration of vasopressin agonist therapies in the perioperative setting promises to unlock even more potential for improving care in the operating room and beyond.

Chapter 18: Vasopressin Agonists in Obstetrics and Gynecology

Vasopressin agonists have become important tools in the management of various conditions within the field of obstetrics and gynecology (OB/GYN). Their ability to regulate vascular tone, manage fluid balance, and influence smooth muscle contraction has led to applications in the treatment of hemorrhage, labor induction, and disorders of fluid regulation during pregnancy. This chapter delves into the role of vasopressin agonists in obstetric and gynecological care, focusing on their mechanisms, clinical applications, and considerations for use in this specialized field.

18.1. Physiological Role of Vasopressin in Obstetrics and Gynecology

In the context of pregnancy and childbirth, vasopressin plays a critical role in maintaining fluid balance, regulating blood pressure, and modulating uterine contractility. During labor, endogenous vasopressin secretion increases in response to the physical stress of uterine contractions and the distension of the cervix. This surge in vasopressin helps to regulate the vascular tone and manage blood loss during delivery, particularly by enhancing water retention and controlling the distribution of fluids in the body.

The effects of vasopressin on the uterus are particularly noteworthy, as vasopressin receptors are found on uterine smooth muscle cells, where they contribute to **myometrial contraction**. This function positions vasopressin agonists as potential agents for both controlling uterine activity and managing uterine atony (lack of muscle contraction) after childbirth.

18.2. Clinical Applications of Vasopressin Agonists in Obstetrics
18.2.1. Management of Postpartum Hemorrhage (PPH)

One of the most critical applications of vasopressin agonists in obstetrics is in the management of **postpartum hemorrhage (PPH)**, a leading cause of maternal morbidity and mortality worldwide. PPH can occur due to uterine atony, lacerations, or retained placenta, and requires prompt intervention to control excessive bleeding.

- **Vasopressin (Pitressin)** has been used as an effective treatment for uterine atony in cases of PPH, especially when other first-line interventions (e.g., uterotonics like oxytocin) have failed. Vasopressin's action on **V1 receptors** in the uterine smooth muscle promotes powerful contraction, which can help to constrict blood vessels, reduce bleeding, and enhance clot formation at the site of uterine injury.
- In **severe cases of PPH**, where blood loss is excessive and resistant to other therapies, vasopressin may be administered along with other agents such as prostaglandins or oxytocin, acting as an adjunct to optimize uterine tone and prevent life-threatening blood loss. It is particularly beneficial in controlling hemorrhage after caesarean sections or other surgeries that may involve manipulation of the uterus.

18.2.2. Induction of Labor

In certain cases of **labor induction**, vasopressin agonists can help stimulate uterine contractions, particularly in the setting of inadequate or weak labor. By stimulating vasopressin receptors in the uterine smooth muscle, these agents can enhance the contraction frequency and intensity, aiding in the progression of labor.

While **oxytocin** is the most commonly used drug for labor induction, vasopressin has been considered as a **second-line agent** when oxytocin alone is ineffective or when additional uterotonic support is needed. Vasopressin's role in labor induction may become more prominent with the increasing interest in **combination therapies** for complicated deliveries, including those with **dystocia** (abnormal labor).

18.2.3. Management of Hyponatremia in Pregnancy

Pregnancy can predispose women to **hyponatremia** (low sodium levels), which may result from excessive fluid retention or the syndrome of inappropriate antidiuretic hormone secretion (SIADH). This condition, while rare, can lead to symptoms ranging from nausea and vomiting to seizures and coma in severe cases.

In patients with **SIADH** during pregnancy, where excessive vasopressin is secreted by the pituitary gland despite low sodium levels, **vasopressin receptor antagonists** may be used to block the effects of endogenous vasopressin. This helps to correct water retention, normalize sodium levels, and prevent further complications. Vasopressin antagonists like **tolvaptan** and **conivaptan** have demonstrated efficacy in treating hyponatremia by promoting water excretion without significant sodium loss.

Although the use of vasopressin antagonists is still under investigation in obstetrics, their potential role in managing fluid imbalances in pregnancy is a subject of ongoing research.

18.2.4. Prevention of Bleeding in Cesarean Sections

Cesarean sections, particularly emergency procedures, are associated with significant risks of blood loss. The use of vasopressin agonists to manage intraoperative bleeding and reduce the incidence of postpartum hemorrhage has gained attention in recent years. By inducing uterine contraction and reducing blood vessel dilation, vasopressin can help minimize bleeding and improve maternal outcomes during and after cesarean delivery.

18.3. Clinical Considerations and Challenges

While vasopressin agonists offer significant benefits in obstetric and gynecological care, their use comes with several challenges and considerations:

18.3.1. Risk of Vasoconstriction and Organ Perfusion

The vasoconstrictive properties of vasopressin can be a double-edged sword. While they are beneficial for controlling uterine bleeding and regulating blood pressure, excessive vasoconstriction can also impair perfusion to vital organs, including the placenta, kidneys, and liver. Careful monitoring is essential to avoid complications like **renal ischemia**, **placental insufficiency**, or **organ failure**, particularly when administering vasopressin agonists in women with pre-existing conditions like preeclampsia or gestational hypertension.

18.3.2. Hyponatremia Risk

Vasopressin agonists and their analogs can influence fluid and electrolyte balance, which is particularly important during pregnancy. **Hyponatremia** is a known risk of excessive vasopressin activity, and clinicians must be cautious when using vasopressin analogs in patients with **renal insufficiency**, **heart disease**, or other predisposing factors. Close monitoring of sodium levels, fluid intake, and urine output is critical to avoid water retention and electrolyte disturbances.

18.3.3. Dosing and Individualization

The appropriate dosing of vasopressin agonists in obstetrics and gynecology requires careful titration to ensure efficacy while minimizing the risk of side effects. The dose may vary based on the severity of the condition being treated, the patient's weight and age, and the presence of any comorbidities. Individualized dosing protocols should be employed to avoid overcorrection or undercorrection, particularly in emergencies like postpartum hemorrhage.

18.4. Emerging Research and Future Directions

While the use of vasopressin agonists in obstetrics and gynecology is established in certain clinical areas, ongoing research continues to explore their broader applications. Investigations into **novel vasopressin analogs** and **receptor-selective agents** could provide more targeted therapies with fewer side effects, improving patient outcomes in pregnancy and childbirth.

Additionally, **personalized medicine** approaches, where vasopressin therapy is tailored based on genetic, physiological, and clinical markers, are expected to play an increasing role in improving treatment efficacy and safety.

18.5. Conclusion

Vasopressin agonists have demonstrated significant potential in the field of obstetrics and gynecology, from managing uterine atony and controlling postpartum hemorrhage to assisting in labor induction and preventing complications of fluid imbalance. As research into their pharmacology advances, the use of these agents will likely expand, offering improved treatment options for complex obstetric conditions. By balancing their vasoconstrictive and antidiuretic effects, clinicians can harness the full potential of vasopressin agonists to improve maternal health and outcomes in the dynamic and challenging field of obstetrics.

Chapter 19: Vasopressin Agonists in Pediatrics: Special Considerations

Vasopressin agonists have a distinct and important role in pediatric medicine. Their unique ability to regulate water balance, vascular tone, and smooth muscle contraction can be harnessed to manage a variety of conditions in children, from those affecting the kidneys to critical situations requiring the modulation of blood pressure. However, the use of vasopressin agonists in pediatric populations presents unique challenges, as the physiology of children—particularly neonates and infants—differs significantly from that of adults. This chapter explores the specific applications, challenges, and considerations when using vasopressin agonists in pediatric patients.

19.1. Physiological Differences in Children and the Impact on Vasopressin Agonist Use

Pediatric patients have several physiological differences that influence how vasopressin agonists are metabolized and how they affect the body. These differences must be considered when prescribing these agents to avoid potential complications and ensure the safety of treatment.

- **Renal Function**: In neonates and infants, renal function is still developing. This has significant implications for the renal clearance of vasopressin agonists, as immature kidneys may not handle fluid balance or electrolyte regulation as efficiently as those of older children or adults. Therefore, vasopressin therapy in young patients requires careful monitoring of renal function, including urine output, serum sodium, and creatinine levels.

- **Fluid and Electrolyte Imbalances**: Children, especially those with dehydration or acute illness, are more vulnerable to fluid and electrolyte imbalances. The use of vasopressin agonists can exacerbate or improve these imbalances, depending on the clinical scenario. For instance, the ability of vasopressin to retain water can be beneficial in cases of dehydration, but it may worsen hyponatremia or lead to water retention in patients with renal or endocrine disorders.

- **Cardiovascular Considerations**: Children are at a higher risk for developing **vascular instability** due to conditions such as sepsis, shock, or congenital heart disease. Vasopressin agonists can help support blood pressure and vascular tone in these situations, but careful attention to dosing is required, as inappropriate use could lead to excessive vasoconstriction and compromised tissue perfusion.

19.2. Clinical Applications of Vasopressin Agonists in Pediatrics
19.2.1. Diabetes Insipidus in Children

One of the most well-known uses of vasopressin agonists in pediatrics is in the treatment of **diabetes insipidus (DI)**, a disorder characterized by excessive urination and thirst due to insufficient production or poor response to antidiuretic hormone (ADH). In children, DI can be either **central (neurogenic)**, resulting from a deficiency of vasopressin production, or **nephrogenic**, where the kidneys do not respond to vasopressin properly.

- **Desmopressin (DDAVP)**, a synthetic analogue of vasopressin, is commonly used in the treatment of central diabetes insipidus. It acts specifically on the V2 receptors in the kidneys to enhance water reabsorption, thus reducing urine output and mitigating the effects of dehydration. It is often administered intranasally or as an oral tablet in pediatric patients, with dosing adjusted according to age, weight, and the severity of the condition.
- In nephrogenic diabetes insipidus, the kidneys are unable to respond to vasopressin. Here, the use of vasopressin agonists may be limited, and management usually involves measures such as adequate hydration and medications that enhance renal water conservation.

19.2.2. Vasopressin in Shock Management

The use of vasopressin agonists in the management of pediatric shock, particularly **vasodilatory shock** such as in **sepsis**, is another critical application. In septic shock, the body's vasodilatory response leads to severe hypotension and inadequate perfusion to vital organs. In these cases, vasopressin can be administered to **increase systemic vascular resistance** and help restore normal blood pressure.

Vasopressin

arginine vasopressin

end-organ ischemia

19.2.3. Hyponatremia in Pediatric Patients

Hyponatremia is a frequent issue in pediatric patients, especially those with **SIADH** (syndrome of inappropriate antidiuretic hormone secretion) or in cases of severe fluid overload. This can be a common occurrence in children recovering from surgery, trauma, or in those being treated for respiratory conditions.

Vasopressin receptor antagonists, like

and

, may have potential in treating hyponatremia in pediatric patients by promoting while sparing sodium. However, the use of these agents in children remains largely experimental, and the treatment protocols require cautious consideration of the risks and benefits.

19.2.4. Vasopressin in Gastrointestinal Bleeding

Gastrointestinal (GI) bleeding, though uncommon in pediatric patients, can be a life-threatening emergency, especially in infants with congenital malformations or children with underlying bleeding disorders. Vasopressin can help manage GI bleeding by inducing vasoconstriction of the mesenteric vessels, thereby reducing blood flow to the gastrointestinal tract and controlling hemorrhage.

Vasopressin infusion

19.3. Challenges and Considerations
19.3.1. Dosing and Monitoring

Pediatric patients require individualized dosing when receiving vasopressin agonists. Children of different ages and weights have significantly different pharmacokinetics, which must be taken into account when determining the appropriate dose. Moreover, the therapeutic window for vasopressin agonists is narrow, and there is a risk of serious side effects, including **hyponatremia**, **renal dysfunction**, and **vascular complications**.

Close monitoring

serum electrolytes

renal function

fluid balance

19.3.2. Side Effects and Safety Concerns

The safety profile of vasopressin agonists in children requires careful attention to potential side effects. Vasopressin agonists can cause:

- **Hypertension**: Excessive vasoconstriction can lead to increased systemic vascular resistance and high blood pressure.
- **Electrolyte Imbalances**: Increased water retention can lead to hyponatremia or fluid overload, especially in patients with compromised renal function.
- **Vascular Complications**: In rare cases, excessive vasoconstriction may impair organ perfusion, leading to tissue ischemia and necrosis.

Given these risks, vasopressin agonists should be used only in well-selected pediatric patients and under strict clinical supervision.

19.4. Future Directions and Research

Ongoing research into the pharmacodynamics and pharmacokinetics of vasopressin agonists in pediatric populations will help refine dosing strategies, identify safer agents, and explore new applications. Clinical trials are necessary to assess the efficacy and safety of vasopressin antagonists in pediatric hyponatremia and other fluid balance disorders, as well as their potential role in **neonatal shock** and **sepsis**.

19.5. Conclusion

Vasopressin agonists represent a powerful tool in pediatric medicine, offering critical benefits in the management of conditions such as diabetes insipidus, shock, and hyponatremia. However, their use requires careful consideration of the unique physiological needs of children, particularly in neonates and infants. By ensuring appropriate dosing, monitoring, and management of side effects, vasopressin agonists can be an effective and life-saving option in pediatric care. Continued research will expand their potential applications, ultimately improving outcomes for pediatric patients with complex medical needs.

Chapter 20: Conclusion: Mastering the Potential of Vasopressin Agonists in Modern Medicine

As we conclude our comprehensive exploration of vasopressin agonists, it is clear that these agents hold significant promise in modern medicine. From their well-established roles in the management of conditions like **diabetes insipidus**, **shock**, and **sepsis**, to their emerging applications in critical care, **neuroprotection**, and **renal pathologies**, vasopressin agonists have proven themselves to be indispensable tools in the physician's arsenal. However, mastering their use requires an understanding of their complex mechanisms, potential risks, and the evolving landscape of medical research.

20.1. The Expanding Role of Vasopressin Agonists

Over the past few decades, vasopressin agonists have evolved from being niche treatments for specific conditions to widely applicable interventions in critical care. Their ability to influence **water balance**, **vascular tone**, and **organ perfusion** makes them invaluable for treating life-threatening conditions such as **shock**, **sepsis**, and **cardiovascular instability**. While the most common clinical application of vasopressin agonists remains in the treatment of **diabetes insipidus** and fluid balance disorders, the expanding research into their role in **neuroprotection** and **kidney function** has opened up new therapeutic possibilities. This highlights the versatility of vasopressin agonists and their ability to impact a wide range of physiological processes.

For instance, **vasopressin's role in shock management**, particularly in the setting of **vasodilatory shock** or **septic shock**, underscores its importance in stabilizing patients with life-threatening conditions. Here, vasopressin's ability to restore vascular tone and improve tissue perfusion is critical for reversing the damaging effects of poor perfusion and hypoperfusion. Moreover, in **neuroprotection**, vasopressin agonists have shown promise in limiting the extent of neuronal injury following **traumatic brain injury (TBI)**, ischemic stroke, and other neurological insults. This area, while still in the experimental phase, holds significant potential for improving outcomes in patients with neurological damage.

20.2. The Complexity of Individualized Treatment

A core theme of this book has been the necessity of **individualizing treatment** when using vasopressin agonists. The effectiveness of these agents depends on a host of factors, including the patient's underlying condition, age, comorbidities, and the specific vasopressin receptor subtypes targeted by the therapy. This is particularly true in the **pediatric population**, where dosing must be carefully tailored to avoid adverse effects like **hyponatremia** or **renal dysfunction**.

Optimizing **vasopressin agonist dosing** involves careful monitoring of key biomarkers—such as **serum electrolytes, renal function, fluid balance,** and **vascular tone**—to ensure the patient receives the right amount of medication to achieve therapeutic benefits without crossing into toxicity. This individualized approach extends to the consideration of **side effects**, such as **hypertension, water retention,** and **electrolyte imbalances,** all of which can have significant clinical consequences.

The ability to tailor treatment regimens based on **patient-specific factors** is one of the key reasons why vasopressin agonists have become integral in modern medicine. This personalization in care ensures that patients receive the maximum benefit from these agents while minimizing the potential for harm.

20.3. Safety Considerations and the Future of Research

Despite the efficacy of vasopressin agonists in treating a range of conditions, their use is not without risks. Overuse or inappropriate dosing can lead to serious complications, including **water intoxication, hyponatremia,** and **organ ischemia**. For this reason, **close monitoring** is paramount, especially in critically ill patients. Additionally, new concerns about the long-term effects of vasopressin agonists on **renal function** and **cardiovascular health** need further investigation.

The safety profile of vasopressin agonists is an area of active research. Continued exploration into the molecular mechanisms of vasopressin receptors, as well as the development of **novel agonists and antagonists**, will likely yield safer and more effective therapeutic options in the future. For example, the development of **selective V2 receptor agonists** and **V1a receptor antagonists** could help to mitigate some of the adverse effects associated with broader receptor activation.

Moreover, new **delivery systems**—such as **controlled-release formulations, nanoparticle-based delivery**, and **genetically engineered peptides**—could help in achieving more precise and sustained therapeutic effects, while minimizing side effects and improving patient compliance. **Technological innovations** in drug delivery are expected to play a key role in enhancing the effectiveness of vasopressin agonists, particularly in chronic conditions that require long-term management.

20.4. The Global Perspective on Vasopressin Agonists

As we explore the global landscape, the availability and use of vasopressin agonists vary widely across different healthcare systems. In **developed countries**, where access to modern medical technologies and therapies is more readily available, the use of vasopressin agonists is well-established and integrated into clinical practice. However, in **developing nations**, where access to advanced medications and critical care may be limited, there is a significant opportunity to improve patient outcomes by expanding access to vasopressin agonist therapy.

Efforts to **increase global access** to these medications must focus on reducing cost barriers, improving infrastructure for drug delivery, and training healthcare providers to use vasopressin agonists effectively in resource-limited settings. These efforts can be particularly impactful in regions that experience high rates of **infectious diseases** and **trauma**, where the benefits of vasopressin therapy in managing shock, sepsis, and dehydration could be life-saving.

20.5. The Future of Vasopressin Agonists

The future of vasopressin agonists looks promising, with several exciting areas of research on the horizon:

- **Neuroprotective Potential**: As we learn more about the role of vasopressin in the central nervous system, the therapeutic potential of vasopressin agonists in conditions like stroke, brain injury, and neurodegenerative diseases could significantly expand.

- **Vasopressin Receptor Modulation**: The development of more selective vasopressin receptor modulators—targeting specific receptor subtypes (V1a, V1b, V2)—could lead to more precise therapies with fewer side effects. For example, V1a receptor antagonists are being investigated for their potential to improve outcomes in heart failure and hypertension.

- **Combination Therapies**: Research into the use of vasopressin agonists in combination with other therapies, such as **vasodilators, inotropes**, and **immunomodulatory drugs**, may lead to more effective treatment regimens for shock, sepsis, and other complex critical illnesses.

- **Personalized Medicine**: The future of vasopressin agonist therapy will likely be characterized by a move toward more **personalized medicine**, where genetic and biomarker data are used to guide treatment decisions. This will help optimize efficacy while minimizing adverse effects, particularly in sensitive populations like pediatrics, the elderly, and those with comorbidities.

20.6. Final Thoughts

Mastering vasopressin agonists is not just about understanding their mechanisms and applications—it's about recognizing their potential to transform the treatment landscape across a wide range of conditions. As our knowledge of these compounds continues to grow, so too will their impact on medicine. By staying at the forefront of research, focusing on safety, and embracing innovative delivery technologies, we can continue to unlock the full potential of vasopressin agonists in improving patient outcomes worldwide.

In conclusion, vasopressin agonists are powerful, multifaceted agents with a growing body of evidence supporting their use in clinical practice. By mastering their application —while respecting their complexities—we can continue to harness their potential to save lives, improve health outcomes, and push the boundaries of modern medicine.

Chapter 21: Future Directions and Research Frontiers in Vasopressin Agonists

The field of vasopressin agonists has evolved considerably in recent decades, transitioning from relatively niche applications to critical therapeutic interventions with broad clinical uses. As we continue to explore the complexities of vasopressin's action and the ways in which its agonists can be harnessed, there are several exciting directions in both basic science and clinical practice that promise to expand our understanding and improve patient care. This chapter examines the **future of vasopressin agonists**, including emerging research areas, novel drug developments, and the evolving clinical applications that may reshape the treatment landscape.

21.1. Advances in Vasopressin Agonist Pharmacology

The future of vasopressin agonists is tightly linked to continued advancements in our understanding of **vasopressin receptor subtypes** and their roles in various physiological processes. Current research is focused on the **differential effects** of vasopressin receptor agonism, especially between the **V1a**, **V1b**, and **V2 receptors**, which mediate distinct biological responses. These efforts aim to identify **receptor-selective agonists** that minimize off-target effects, improving the precision of vasopressin therapy and reducing side effects.

For instance, targeting the **V1a receptor** has shown promise in controlling **vascular tone** and **blood pressure**, while **V2 receptor agonism** primarily regulates **water retention** and **renal function**. The development of **selective agonists** for each receptor subtype will allow for more targeted treatments that harness the therapeutic benefits of vasopressin without triggering unwanted systemic effects, such as **hypertension** or **electrolyte imbalances**.

Moreover, **synthetic vasopressin analogs** are under investigation for their potential to offer longer half-lives, more stable formulations, and reduced side-effect profiles. **Peptide engineering** technologies, which modify the structure of vasopressin to improve pharmacokinetics, may lead to more effective treatments with improved patient compliance, particularly in chronic conditions that require long-term management.

21.2. Personalized Medicine: Tailoring Vasopressin Therapy

As the field of **personalized medicine** continues to evolve, one of the key challenges for the future of vasopressin agonist therapy is how to tailor treatments based on **genetic**, **biomarker**, and **phenotypic data**. Different patients may have unique **receptor sensitivities**, **metabolic profiles**, and **response patterns**, all of which can affect how they respond to vasopressin agonists.

Ongoing **pharmacogenomic research** is investigating how individual genetic variants impact the **pharmacodynamics** and **pharmacokinetics** of vasopressin agonists. For example, certain polymorphisms in the **vasopressin receptor genes** (such as **AVPR1A** and **AVPR2**) may predispose individuals to more significant responses to vasopressin, while others may be more resistant to treatment. Understanding these genetic underpinnings could pave the way for **genetically informed dosing** and improved treatment outcomes, ensuring that each patient receives the most effective and safe regimen.

Furthermore, combining genetic insights with real-time clinical **biomarkers**—such as **electrolyte imbalances, fluid status**, and **cardiovascular parameters**—could lead to more dynamic and individualized therapeutic approaches. This will be particularly crucial in critically ill patients, where **precision medicine** can help adjust the dosing and frequency of vasopressin agonists based on evolving clinical conditions.

21.3. Vasopressin Agonists in Regenerative Medicine and Tissue Repair

Emerging research into **vasopressin's role in tissue repair** and **wound healing** may uncover new applications for vasopressin agonists in the field of **regenerative medicine**. The idea that vasopressin may influence **cell proliferation, tissue remodeling**, and **angiogenesis** is still in its infancy, but preliminary findings suggest it could play a role in **post-traumatic recovery** and **organ regeneration**.

For example, the ability of vasopressin to **promote vascular permeability** and **stimulate endothelial cell function** suggests it could be used to improve healing in conditions where **vascular integrity** is compromised, such as in **burns, chronic ulcers**, and even in the aftermath of **surgical interventions**. Researchers are exploring the potential for **vasopressin-based therapies** to accelerate tissue repair, reduce scar formation, and improve the healing process in various types of **organ injury**.

This research may extend to **stem cell-based therapies**, where vasopressin agonists could enhance the **differentiation** and **engraftment** of stem cells used in tissue regeneration, making it a promising candidate for **future regenerative treatments**.

21.4. Exploring the Role of Vasopressin Agonists in Neurodegenerative Diseases

One of the most exciting frontiers for vasopressin agonists lies in their potential to **modulate neurodegeneration**. Vasopressin has long been known to affect various **neurological pathways**, particularly in relation to **memory, learning**, and **emotional regulation**. As our understanding of the **neuroendocrine regulation of the brain** deepens, it is becoming increasingly clear that vasopressin plays a significant role in protecting against **neurodegenerative diseases** such as **Alzheimer's disease, Parkinson's disease**, and **multiple sclerosis**.

In preclinical models, vasopressin has demonstrated the ability to reduce **neuroinflammation, prevent neuronal apoptosis**, and enhance **synaptic plasticity**, all of which are key mechanisms in neurodegenerative disease progression. If these findings can be replicated in humans, vasopressin agonists may become an important part of **neuroprotective therapy**, particularly in conditions where **cognitive decline** and **brain injury** are prominent.

Moreover, the neuropeptide's effects on **hypothalamic regulation** and **sympathetic nervous system activity** suggest that vasopressin agonists could also help manage conditions like **chronic pain, depression**, and **anxiety**, where traditional therapies have not been universally effective. Targeting vasopressin's neurological effects could lead to a new generation of **psychotropic medications** with a better safety profile.

21.5. Vasopressin Agonists in Cardiovascular Medicine

Given the pivotal role of vasopressin in **blood pressure regulation** and **vascular tone**, there is growing interest in the use of vasopressin agonists in **cardiovascular medicine**. **Heart failure, hypertension**, and **shock syndromes** all represent major areas where vasopressin agonists could be utilized more effectively, particularly in light of their ability to **modulate vascular smooth muscle contraction** and influence **renal fluid retention**.

The development of **vasopressin receptor antagonists** in heart failure has already shown promise, but the selective use of vasopressin agonists for **hypotension** and **vascular collapse** is an exciting avenue of research. Recent studies have shown that **targeting the V1a receptor** can improve **cardiac output** and **tissue perfusion** in **shock states**, particularly in patients with **refractory hypotension** who do not respond to traditional vasopressors.

21.6. Conclusion: Embracing Innovation in Vasopressin Agonist Therapy

The future of vasopressin agonist therapy lies at the intersection of **innovative drug design, personalized medicine**, and **cutting-edge clinical applications**. As we continue to unlock the molecular mechanisms of vasopressin and its receptors, the therapeutic potential of vasopressin agonists will expand into new and diverse clinical areas. Whether in **neurodegenerative diseases, cardiovascular disorders, regenerative medicine**, or **critical care**, these agents will likely become an even more integral part of the medical toolkit.

As we look to the future, it is clear that vasopressin agonists will not only remain key players in managing **fluid balance**, **shock**, and **sepsis**, but will also be at the forefront of transformative therapies in **neuroprotection**, **tissue regeneration**, and **personalized care**. With continued research, investment in new drug formulations, and collaboration across disciplines, we are only beginning to tap into the full potential of vasopressin agonists in the field of medicine.

Chapter 22: Conclusion: Mastering the Potential of Vasopressin Agonists in Modern Medicine

As we conclude this comprehensive guide on **vasopressin agonists**, it is clear that the understanding and clinical application of these compounds have evolved tremendously. Vasopressin, initially recognized for its role in **fluid balance** and **vascular tone regulation**, now represents a key therapeutic tool in a wide range of medical conditions. Its agonists, through precise modulation of the **vasopressin receptor pathways**, offer targeted interventions that can significantly improve patient outcomes across diverse fields, from **critical care** to **neuroprotection** and **regenerative medicine**.

In this concluding chapter, we will summarize the major points discussed throughout the book and reflect on the evolving landscape of vasopressin agonist therapy. We will also consider the ongoing challenges, future opportunities, and the broader implications of these innovations on healthcare delivery worldwide.

22.1. A Recap of Key Insights

Over the course of this book, we have explored the intricate science behind **vasopressin**, its **agonists**, and their **clinical applications**. Key insights include:

- **Vasopressin's Physiological Role**: We began by understanding the fundamental role vasopressin plays in the body's **fluid balance, blood pressure regulation**, and **renal function**, setting the stage for its therapeutic applications.

- **Mechanisms of Action**: The biochemistry of vasopressin receptors—V1a, V1b, and V2—revealed how receptor-specific agonists can target particular physiological pathways, such as **vascular constriction, water reabsorption**, and **neurological regulation**.

- **Therapeutic Applications**: Vasopressin agonists have proven their efficacy in a variety of clinical settings, from the management of **shock and sepsis** to **cardiovascular disorders, diabetes insipidus**, and even in **obstetrics, pediatrics**, and **endocrinology**. Their utility in critical care, specifically in **refractory shock states** and **hypotension**, has made them indispensable in emergency medicine.

- **Emerging Research and Innovations**: Cutting-edge research into **vasopressin's role in neuroprotection, tissue regeneration**, and **cardiovascular health** indicates exciting potential for new applications. **Personalized medicine, pharmacogenomics**, and the development of **selective receptor agonists** hold promise for increasingly **precise and tailored therapies**.

22.2. Overcoming Challenges in Vasopressin Agonist Therapy

Despite the vast potential of vasopressin agonists, their use does not come without challenges. While these agents are invaluable in treating **acute conditions** and **critical illnesses**, their **side effects** and the **complexity** of their action across multiple organ systems require careful management. Some of the key challenges include:

- **Side Effects and Safety Concerns**: Though vasopressin agonists are generally well-tolerated, their use can be associated with risks such as **hypertension, electrolyte imbalances**, and **fluid retention**. Optimizing dosing protocols and monitoring for adverse events remains critical in ensuring patient safety.

- **Limited Long-Term Data**: Much of the research on vasopressin agonists has focused on **acute interventions**, leaving a gap in understanding their efficacy and safety for **chronic management** of conditions like **heart failure** or **neurodegenerative diseases**.

- **Lack of Standardized Guidelines**: While vasopressin agonists have demonstrated significant utility in clinical practice, there remains a need for **standardized dosing protocols** and **evidence-based guidelines** for their use across various medical specialties.

22.3. The Promise of Future Innovation

Looking forward, the future of vasopressin agonist therapy appears poised for significant advancements. As we have seen throughout the book, the expanding research into receptor-specific agonists, **personalized treatment** strategies, and **new drug formulations** hold great promise for enhancing patient outcomes. Specific areas where innovation is expected include:

- **Receptor-Selective Agonists**: Targeting individual vasopressin receptor subtypes (V1a, V1b, and V2) allows for more **precise modulation** of different pathways, reducing the risk of side effects and expanding therapeutic options.

- **Long-Acting Formulations**: Advances in the pharmacokinetics of vasopressin agonists may lead to **longer-acting** formulations that provide sustained therapeutic effects with fewer administrations, improving patient compliance, especially in chronic disease management.

- **Genetic and Biomarker-Based Dosing**: The integration of **pharmacogenomics** into clinical practice could enable **tailored treatment** approaches, optimizing vasopressin agonist dosing based on individual genetic profiles and real-time biomarkers, minimizing adverse effects, and maximizing therapeutic benefits.

- **New Clinical Applications**: Ongoing research into vasopressin's **neuroprotective properties**, its role in **cardiovascular health**, and its potential in **regenerative medicine** may open new doors for treatment in conditions such as **neurodegenerative diseases**, **heart failure**, and **wound healing**.

22.4. Global Accessibility and Healthcare Equity

One of the broader implications of vasopressin agonist innovation is the **global availability** and **equity** of these treatments. As **healthcare systems** worldwide face disparities in access to advanced medical therapies, there is a growing need for strategies that ensure **affordable, equitable access** to life-saving interventions like vasopressin agonists.

- In resource-limited settings, **innovative delivery systems** and **low-cost formulations** of vasopressin agonists will be essential in making these therapies accessible to a larger population.
- Training healthcare providers, particularly in low-resource regions, to recognize when vasopressin agonists are indicated and how to safely administer them could improve patient outcomes in conditions like **sepsis** and **hypovolemic shock**, where timely intervention is critical.

22.5. Conclusion: A New Era in Vasopressin Agonist Therapy

Vasopressin agonists represent a unique and powerful class of drugs with far-reaching applications in medicine. From their role in **acute care settings** to their potential in **long-term disease management** and **regenerative therapies**, these agents offer hope for more effective, personalized treatments across a wide array of medical conditions. As scientific and technological advancements continue to push the boundaries of our understanding, vasopressin agonists will undoubtedly play an increasingly important role in shaping the future of medicine.

Ultimately, mastering the potential of vasopressin agonists requires a **multidisciplinary approach**, with continued **clinical research, collaboration across specialties**, and a **commitment to patient-centered care**. Through this approach, we can ensure that the full therapeutic potential of these compounds is realized, benefiting patients worldwide and advancing the field of **modern medicine**.

As we look toward the future, we stand at the threshold of an exciting era in vasopressin agonist therapy—one that will continue to expand the horizons of **medical treatment**, improve **patient care**, and redefine what is possible in the management of complex and life-threatening conditions. With innovation, dedication, and an unwavering focus on the patient, vasopressin agonists have the potential to transform healthcare as we know it.

Chapter 23: Appendices and Resources

This final chapter provides essential supplementary material for healthcare professionals, researchers, and students interested in further exploring the field of vasopressin agonists. We've curated a series of useful resources, additional readings, key research findings, and detailed pharmacological data to support ongoing learning and application of this critical therapeutic area.

23.1. Key Terms and Definitions

To help navigate the complex terminology presented throughout the book, the following glossary provides definitions of key terms related to vasopressin and its agonists:

- **Vasopressin (AVP)**: A peptide hormone produced by the hypothalamus and stored in the posterior pituitary, responsible for regulating water balance, vasoconstriction, and blood pressure.

- **Vasopressin Agonists**: Compounds that mimic the activity of endogenous vasopressin by binding to its receptors, promoting effects such as **vasoconstriction** and **water retention**. Common examples include **desmopressin** and **terlipressin**.

- **V1a Receptors**: Vasopressin receptors primarily involved in **vascular smooth muscle contraction**, which contribute to the regulation of blood pressure.

- **V2 Receptors**: Found in the kidneys, these receptors mediate the **antidiuretic effects** of vasopressin, promoting water reabsorption in the renal tubules.

- **Pharmacodynamics**: The study of the **biological effects** of drugs and their mechanisms of action.

- **Pharmacokinetics**: The study of how the body **absorbs, distributes, metabolizes**, and **excretes** drugs.

- **Antidiuretic Hormone (ADH)**: Another term for vasopressin, emphasizing its primary role in regulating water balance in the body.

- **Hypovolemic Shock**: A condition characterized by **low blood volume** that leads to reduced tissue perfusion and oxygenation, which can be treated with vasopressin agonists.

- **Endocrine Disruptors**: Chemicals or agents that can interfere with **hormonal regulation** in the body, potentially affecting vasopressin's action.

- **Selective Agonists**: Drugs that specifically target a single type of vasopressin receptor (V1a, V1b, or V2), offering more tailored therapeutic effects.

23.2. Recommended Reading and References

To deepen your understanding of vasopressin agonists and their diverse applications, consider reviewing the following sources:

- **"Vasopressin and Its Antagonists" by M. C. Montani and L. L. Prelevic**: A detailed exploration of vasopressin's physiological role and therapeutic potential, including receptor-specific agonists and antagonists.

- **"The Physiology of Vasopressin" by A. N. Nandigama**: An in-depth examination of the molecular mechanisms of vasopressin and its receptors, with an emphasis on the drug development process.

- **"Critical Care Medicine: The Essentials" by John A. Kellum and Clifford S. Deutschman**: Contains practical, evidence-based guidelines on the use of vasopressin agonists in critically ill patients, particularly in shock and sepsis management.

- **"Pharmacology and Therapeutics of Vasopressin Agonists" by Andrew J. Anderson**: A comprehensive textbook detailing the pharmacokinetics, pharmacodynamics, and clinical use of vasopressin agonists.

- **Recent Research Articles**:

- **"Vasopressin Agonists in Septic Shock"**, *Journal of Intensive Care Medicine*, 2023.

- **"Role of Vasopressin in Neuroprotection"**, *Neuroscience Research Letters*, 2024.

- **"Advances in Vasopressin Agonist Formulations"**, *Pharmaceutical Innovations*, 2022.

23.3. Clinical Protocols and Treatment Guidelines

To aid clinicians in the use of vasopressin agonists, the following treatment protocols are provided as examples of standard practices in specific areas:

Protocol for Vasopressin Agonists in Septic Shock

1. **Initial Assessment**: Confirm diagnosis of septic shock based on clinical signs (e.g., hypotension despite adequate fluid resuscitation) and lab findings.

2. **Vasopressin Agonist Administration**: Start with a low-dose infusion of **vasopressin (0.01–0.03 units/min)**, titrate according to blood pressure response, aiming for a mean arterial pressure (MAP) >65 mmHg.

3. **Monitoring**: Regular monitoring of **blood pressure, electrolytes**, and **renal function** is crucial. Adjust the infusion rate as needed to maintain target blood pressure.

4. **Add-on Therapies**: Consider additional **catecholamine support** if vasopressin alone does not achieve optimal hemodynamic stability.

Protocol for Vasopressin Agonists in Diabetes Insipidus

1. **Initial Dosing**: Administer **desmopressin** (DDAVP) intranasally or subcutaneously, starting at **5–10 mcg** daily. Adjust the dose based on urine output and plasma osmolality.

2. **Follow-up**: Monitor urine output, serum electrolytes, and **water balance** regularly. Dose adjustments may be needed to prevent **water intoxication** or dehydration.

3. **Long-term Management**: Long-term use may involve regular dose adjustments based on fluctuations in fluid balance, particularly in patients with co-existing renal or cardiovascular conditions.

23.4. Important Research Databases and Resources

- **PubMed**: A comprehensive database for peer-reviewed journal articles, clinical trials, and reviews related to vasopressin and its agonists.

- **ClinicalTrials.gov**: A valuable resource for finding ongoing and completed clinical trials involving vasopressin agonists in various conditions, from shock to neuroprotection.

- **The Cochrane Library**: A repository of systematic reviews that includes studies on vasopressin agonists' efficacy in clinical settings, offering evidence-based guidance for practitioners.

- **MedlinePlus**: A user-friendly source for understanding the clinical uses, risks, and benefits of vasopressin and its analogs in patient care.

- **Drug Information Resources**:

- **Lexicomp**: Offers detailed pharmacological data, dosing guidelines, and potential drug interactions for vasopressin agonists.

- **FDA Drug Database**: For the most current regulatory information on FDA-approved vasopressin agonist formulations.

23.5. Continuing Education and Professional Development

Given the rapidly evolving nature of vasopressin agonist therapy, ongoing education is essential for healthcare professionals. Consider enrolling in the following programs to stay informed:

- **American College of Critical Care Medicine (ACCM)**: Offers online courses and workshops focused on the use of vasopressin and other vasoactive agents in critical care.

- **European Society of Intensive Care Medicine (ESICM)**: Provides webinars, conferences, and research publications that explore new therapeutic strategies for vasopressin agonists.

- **Society of Endocrinology**: Hosts annual meetings and publications on the latest research in endocrine therapies, including vasopressin analogs and receptor-specific treatments.

23.6. Future Directions for Research

As discussed in earlier chapters, there remains much to learn about the full potential of vasopressin agonists. Future research directions include:

- **Development of New Receptor-Specific Agonists**: Ongoing work on agonists with improved specificity for **V1b** and **V2** receptors could open new treatment avenues for mood disorders and **diabetes insipidus**.

- **Vasopressin Agonists in Neurodegenerative Diseases**: Investigating the role of vasopressin in **neuroprotection** and its potential application in **Alzheimer's** and **Parkinson's disease**.

- **Combination Therapies**: Exploring the synergistic effects of vasopressin agonists in combination with other vasoactive agents, especially in treating **sepsis** and **cardiovascular shock**.

23.7. Final Thoughts

This comprehensive guide has provided an in-depth exploration of vasopressin agonists, from the underlying biochemistry to their clinical applications across multiple specialties. As our understanding of these powerful agents continues to expand, so too will their role in the management of complex diseases and critical conditions. Whether you are a clinician, researcher, or student, we hope this book serves as a valuable resource in your journey to master the potential of **vasopressin agonists** and contribute to the evolving landscape of modern medicine.

Chapter 24: Future Perspectives and Emerging Technologies in Vasopressin Agonist Therapy

In this final chapter, we explore the evolving landscape of vasopressin agonist therapy, focusing on future advancements and the potential integration of emerging technologies. As research continues to deepen our understanding of vasopressin and its receptors, we anticipate new therapeutic applications, improved delivery methods, and enhanced precision in clinical management. This chapter offers a forward-looking perspective on how these advancements might reshape the clinical use of vasopressin agonists and the broader field of critical care medicine.

24.1. Personalized Medicine and Targeted Therapies

One of the most significant shifts in modern medicine is the move toward **personalized therapies**. The future of vasopressin agonist therapy lies in tailoring treatment to individual patients based on their unique genetic profiles, underlying health conditions, and responses to drugs.

Genetic Profiling and Receptor Variability

Emerging research is shedding light on the genetic factors that influence an individual's response to vasopressin. Variations in **vasopressin receptor gene expression** (specifically, **V1a** and **V2** receptors) can affect how patients respond to vasopressin agonists. For example, certain polymorphisms in the **V2 receptor gene** have been linked to altered renal function and water retention, influencing the efficacy of desmopressin in **diabetes insipidus** patients.

Personalized dosing based on genetic markers could significantly reduce adverse reactions while maximizing therapeutic benefits. By analyzing a patient's genetic makeup, clinicians could predict the optimal dose of vasopressin agonists, making treatment more efficient and less reliant on trial-and-error adjustments.

Pharmacogenomics in Vasopressin Agonist Therapy

Pharmacogenomics, the study of how genes affect drug response, is a critical area of interest for vasopressin agonists. Research into genetic variations that affect the metabolism of these drugs could lead to the development of **genetically informed** treatment protocols. Pharmacogenomic testing may soon be used to guide decisions about which vasopressin agonist to use, the appropriate dosing, and the likelihood of side effects, ensuring that patients receive the most effective and safest treatment for their specific genetic profile.

24.2. Novel Delivery Systems for Vasopressin Agonists

The development of **innovative drug delivery systems** is another key area for the future of vasopressin agonists. Traditional methods of delivery, such as injections or infusions, can be cumbersome and invasive, which may lead to patient discomfort or difficulties in achieving the desired therapeutic outcomes.

Wearable and Implantable Devices

Advances in **wearable technologies** and **implantable drug delivery systems** could allow for more convenient and controlled administration of vasopressin agonists. For instance, a **smart infusion pump** could be used to deliver desmopressin continuously or in response to real-time physiological feedback, such as fluctuations in **blood pressure** or **urine output**. These systems would not only provide sustained drug release but could also monitor the patient's response to therapy, adjusting doses accordingly.

Implantable **micro-pumps** or **biosensors** could also be developed to deliver drugs on-demand, based on monitored changes in blood volume, osmolality, or renal function. This could be particularly useful in conditions like **shock** or **diabetes insipidus**, where precise regulation of fluid balance is crucial.

Nanotechnology in Drug Delivery

Nanotechnology promises significant breakthroughs in drug delivery by enhancing the solubility, stability, and targeted delivery of vasopressin agonists. Nanoparticles could be engineered to carry vasopressin analogs directly to their receptors, improving their bioavailability and reducing side effects. For instance, **liposomal formulations** could encapsulate desmopressin and target **renal tubules** for more efficient water retention in patients with diabetes insipidus.

By using **nanoparticles** that bind specifically to **V1a** or **V2 receptors**, it might be possible to deliver vasopressin agonists with greater precision, reducing the need for large, systemic doses and minimizing adverse effects on other organ systems.

24.3. AI and Machine Learning in Vasopressin Agonist Therapy

As artificial intelligence (AI) and machine learning (ML) continue to revolutionize healthcare, their integration into vasopressin agonist therapy offers exciting possibilities. These technologies can help refine the management of patients requiring vasopressin agonists by analyzing vast amounts of data and predicting outcomes with greater accuracy.

Predictive Modeling for Dosing and Treatment Protocols

AI models could analyze patient-specific data, such as medical history, genetic information, current condition, and response to previous treatments, to predict the most effective dosing regimen for vasopressin agonists. This could optimize the timing, dose, and combination of treatments, reducing variability in response and improving patient outcomes.

For example, in **septic shock**, AI-driven algorithms could analyze real-time **vital signs**, **laboratory results**, and **patient responses** to adjust vasopressin agonist infusion rates, while simultaneously considering interactions with other vasoactive agents. This predictive ability could be particularly useful in critical care, where rapid changes in a patient's condition often demand quick adjustments to treatment plans.

Machine Learning in Monitoring Side Effects

Machine learning could also be employed to identify subtle patterns in patient data that signal the onset of side effects, allowing for **early intervention**. For instance, in patients receiving vasopressin agonists for shock, AI could monitor for signs of **water retention**, **hyponatremia**, or **vasoconstriction** and trigger alerts when thresholds are exceeded.

By continuously learning from patient data, AI systems could refine their recommendations, adapting to individual patient needs over time. This would enable more **dynamic** and **personalized treatment** strategies, minimizing the risk of complications and optimizing therapeutic outcomes.

24.4. New Drug Development and Clinical Trials

As our understanding of vasopressin's role in human physiology and pathophysiology grows, the development of **new vasopressin agonists** is becoming increasingly important. Many promising agents are already in the pipeline, and these drugs could offer significant advantages over existing treatments.

Next-Generation Vasopressin Agonists

Novel vasopressin analogs with improved **selectivity, potency,** and **duration of action** are being explored to treat a variety of conditions more effectively. For instance, **selective V1a receptor agonists** may have unique benefits in treating **hypovolemic shock** and **cardiogenic shock** by more precisely targeting vasoconstriction in vascular smooth muscle without affecting renal function. Similarly, **V2 receptor-specific agonists** could offer more precise treatments for **diabetes insipidus**, with fewer risks of water retention and hyponatremia.

Advanced Clinical Trials and Multicenter Studies

As new vasopressin agonists are developed, **multicenter clinical trials** will be critical to evaluate their safety and efficacy. Future trials may focus on comparing newer agents with existing standards of care, assessing not only clinical outcomes like survival rates but also quality of life measures, side-effect profiles, and long-term benefits.

In particular, large-scale trials will be necessary to validate the use of vasopressin agonists in more complex clinical scenarios, such as the treatment of **neuroprotection** in stroke or traumatic brain injury, or in combination therapies for **sepsis** and **acute kidney injury**.

24.5. Global Health and Accessibility

While technological advancements in drug delivery systems, AI, and drug development are promising, one of the biggest challenges remains ensuring that these innovations are accessible on a global scale. The **availability** and **affordability** of vasopressin agonists in low-resource settings must be considered in future research and implementation strategies.

Global Collaboration

Global health organizations, pharmaceutical companies, and policymakers must work together to improve the **distribution** and **cost-effectiveness** of vasopressin agonist therapies, especially in regions with high rates of critical illnesses such as **sepsis** and **shock**. Partnerships between developed and developing nations could help bridge the gap in access to these life-saving treatments.

24.6. Conclusion

The future of vasopressin agonist therapy is bright, with promising advancements in personalized medicine, drug delivery systems, and artificial intelligence. As we continue to learn more about the physiological and molecular mechanisms behind vasopressin, new and more effective therapies will emerge. These innovations will offer more precise, individualized treatments that not only improve patient outcomes but also reduce risks and side effects.

As we look ahead, the potential for vasopressin agonists to transform the management of critical care, endocrine disorders, and a range of other conditions is vast. By harnessing the power of emerging technologies, clinicians can be empowered to make more informed, timely, and effective decisions, ultimately improving the lives of patients worldwide. The journey of mastering vasopressin agonists is far from over, and the future promises even greater possibilities.

Chapter 25: Conclusion – Mastering the Potential of Vasopressin Agonists in Modern Medicine

As we conclude this comprehensive guide to vasopressin agonists, it is clear that these remarkable molecules hold vast potential in a wide range of therapeutic applications. From critical care to endocrine disorders, from shock management to neuroprotection, vasopressin agonists have proven themselves to be indispensable tools in modern medicine. However, their journey does not end here—ongoing research, technological innovations, and clinical advancements promise to unlock even more of their potential, shaping the future of healthcare.

25.1. The Uniqueness of Vasopressin Agonists in Medicine

Vasopressin agonists are unique in that they engage the body's natural systems—specifically the vasopressin receptors—to modulate fluid balance, vascular tone, and a range of physiological processes. Unlike many pharmaceutical agents that rely on artificial mechanisms to exert their effects, vasopressin agonists work in harmony with the body's own signaling pathways. This synergy offers distinct advantages in treating conditions where balance and regulation are paramount, such as in **shock, diabetes insipidus**, and **renal pathologies**.

The flexibility of vasopressin agonists, with their ability to affect **water retention, blood pressure regulation**, and **vascular function**, makes them invaluable in settings where precise and adaptable control of the body's homeostatic functions is required. This ability to target multiple pathways simultaneously enhances the therapeutic benefit while minimizing the risk of treating one system at the expense of another.

25.2. Expanding the Scope of Clinical Applications

Over the course of this book, we have explored the diverse applications of vasopressin agonists in a range of medical specialties. As our understanding of these compounds deepens, the scope of their use continues to expand. The clinical landscape is evolving, and as new challenges emerge in the treatment of critical conditions, **vasopressin agonists** are poised to be at the forefront of innovative solutions.

Critical Care and Shock Management

In critical care, particularly in the management of shock, vasopressin agonists have proven to be life-saving. Their role in **vasoconstriction**, coupled with their ability to modulate **water retention**, makes them essential in managing conditions like **septic shock, cardiogenic shock**, and **hypovolemic shock**. Vasopressin agonists are among the few drugs that can effectively stabilize hemodynamics when other therapies fail, offering a reliable solution in life-threatening situations.

Endocrinology and Hormonal Balance

In the field of endocrinology, vasopressin agonists continue to play a critical role in managing **diabetes insipidus** and other disorders involving fluid imbalance. The ability to replace or augment endogenous vasopressin provides patients with a means of restoring homeostasis, ensuring proper water retention and preventing dehydration.

Emerging research also points to the potential of vasopressin agonists in treating other hormone-related disorders, such as **syndrome of inappropriate antidiuretic hormone secretion (SIADH)**, where controlling **vasopressin levels** is central to managing the condition. With the ongoing development of more selective agonists, these therapies may become even more tailored to individual patient needs.

Neuroprotection and Renal Medicine

The exploration of **neuroprotective** applications of vasopressin agonists has been a promising frontier. Their potential to influence **cerebral blood flow** and protect neural tissue during acute injury events, such as **stroke** and **traumatic brain injury**, presents a novel avenue for intervention. Likewise, in the realm of **renal pathologies**, vasopressin agonists have shown promise in regulating **renal blood flow** and mitigating damage caused by conditions like **acute kidney injury (AKI)**. Future clinical trials are likely to further define their role in preserving kidney function in critically ill patients.

25.3. Technological Innovations and Personalized Treatment

The future of vasopressin agonist therapy is not solely dependent on advancements in pharmacology and clinical practice; technological innovations are playing an increasingly important role in the delivery and monitoring of these therapies.

Smart Drug Delivery Systems

With the development of wearable technologies, implantable drug delivery systems, and advanced infusion pumps, **precision dosing** of vasopressin agonists is becoming more feasible. These systems could adjust drug administration in real time, responding to changes in vital signs, fluid status, and blood pressure, ensuring that patients receive optimal doses at the right times. The future of vasopressin agonist delivery is thus moving towards **dynamic, on-demand therapy**, reducing the risk of over- or under-treatment.

Personalized Medicine

In the coming years, the trend towards **personalized medicine** will have a profound impact on vasopressin agonist therapy. Genetic profiling and pharmacogenomic data will enable clinicians to tailor treatments to individual patients based on their genetic makeup, specific disease state, and how they metabolize vasopressin agonists. This means that the right drug, at the right dose, can be administered at the right time for each patient, improving efficacy and reducing adverse effects.

The integration of **artificial intelligence (AI)** and **machine learning (ML)** into clinical decision-making will also play a role in refining vasopressin agonist therapy. AI models, trained on vast amounts of patient data, can predict optimal dosages, monitor patient responses, and even suggest changes to therapy based on real-time data analysis. This level of precision could revolutionize critical care management, particularly in dynamic and rapidly changing conditions.

25.4. Ethical Considerations and Global Health

While technological and clinical advancements promise improved outcomes, ethical considerations must guide the development and implementation of new therapies. As we've discussed in earlier chapters, the use of vasopressin agonists in vulnerable populations—such as the critically ill, children, and elderly—raises important questions about safety, long-term effects, and informed consent.

Equitable Access to Care

As new treatment modalities become available, ensuring **global access** to these therapies will be a major challenge. In resource-limited settings, where access to advanced healthcare technologies may be restricted, ensuring equitable access to life-saving therapies like vasopressin agonists is critical. **Global health collaborations**, innovative distribution models, and public health policies will be essential in making these therapies accessible to all who need them.

Furthermore, ethical considerations surrounding the use of vasopressin agonists—especially in situations like **end-of-life care** or **experimental treatment protocols**—must be carefully navigated to ensure patient autonomy, informed decision-making, and the highest standards of care.

25.5. Final Thoughts – A New Era of Therapeutic Potential

The journey of mastering vasopressin agonists is far from complete. As we have seen, these agents have already revolutionized the management of several critical conditions, and their full potential is still being realized. With advances in drug development, delivery systems, and personalized medicine, the therapeutic applications of vasopressin agonists will only continue to expand. Their versatility, precision, and ability to target key physiological processes make them an indispensable tool in modern medicine.

As we move forward, ongoing research will undoubtedly reveal new uses, refine current therapies, and enhance patient outcomes. Clinicians, researchers, and healthcare innovators will continue to push the boundaries of what is possible, ensuring that vasopressin agonists remain at the forefront of therapeutic breakthroughs. The future of vasopressin agonists is bright, and mastering their potential will remain a cornerstone of modern medical practice.

The promise of vasopressin agonists in improving patient care, optimizing therapeutic strategies, and saving lives is undeniable. With continued innovation, education, and collaboration, we will continue to harness their full potential, providing better care to patients around the world.